Minut

kept by th

War Committee of the Covenanters

in the

Stewartry of Kirkcudbright.

in the Years 1640 and 1641.

Kirkcudbright:
Printed and Published by J. Nicholson.
M.dccc.lv.

TO

Sir David Maxwell, of Cardoness, Baronet,

IN ACKNOWLEDGEMENT OF

HIS KINDNESS IN PERMITTING THE PUBLICATION OF THIS BOOK,

THE MANUSCRIPT OF WHICH HAD BEEN CAREFULLY PRESERVED IN THE

CHARTER CHEST AT CARDONESS;

AND

IN TESTIMONY OF

THE MOST RESPECTFUL ESTEEM FOR HIMSELF,

THIS VOLUME

IS HUMBLY INSCRIBED

BY

HIS OBLIGED AND MOST OBEDIENT SERVANT,

J. NICHOLSON.

CONTENTS.

CONTENTS OF APPENDIX.

PREFACE.

IN laying the following sheets before the public, the editor finds it necessary to make a few prefactory remarks, in order to connect them with that period of Scottish History which they are intended to illustrate.

The conversion of the people of Scotland from Romanism, was effected under very different circumstances, and by very different instruments, from those by which the same results were obtaind in England. In the latter country, the upper and middle classes had already become wealthy, and felt their property to be tolerably secure, through the establishment of wise and equitable laws; and hence had issued a degree of refinement of manners, and submission to constituted authority, to which their northern neighbours were as yet comparatively strangers.

Owing, mainly perhaps, to a series of short and unfortunate reigns, the Scottish monarchs had never been powerful enough to restrain the aristocracy within the limits of the law, while in England, at this period, to the influence of the law was added the power of one of the most vigorous soverigns that ever swayed the sceptre of that kingdom—and hence the wide difference of condition under which the reformed religion found general admission into the respective countries.

Long prior to the reign of Henry the eighth, however, the reformed doctrines had been taught in England with more or less success, in despite of all opposition from the Romish clergy and the monarch combined; so that it was not through the sweeping edicts of Henry alone, as has been unjustly maintained by the adherents of the ancient faith, that the new doctrines were so universally embraced. The people were evidently in a great measure prepared for the change—but such was the power of the clergy over the people, and so mercilessly did they exercise it, that it was not until Royalty itself found it convenient to repudiate the whole system, and to take the sole management of ecclesiastical affairs into its own hands, that the people at large durst openly venture to adopt that purer version of Christianity for which their minds had been previously prepared.

There is evidently a something in the Anglo-Saxon race, stubbornly adverse to whatever would impede the free march of its onward course. Popery had, it is true, been deeply rooted in the soil, but the genius of the people, with whom it had to deal, was too elastic to be much longer tamely held within its benumbing grasp. This Henry doubtless well knew, else he had certainly paused ere he ventured to substitute himself in the room of a supremacy to which, ostensibly at least, every other potentate in Europe had hitherto yielded submissive obedience. It was, then, through a strong bias in the minds of the people in favours of the Reformation, at that time convulsing the great heart of Germany; together with the impelling force of his own political views, as well as that of his unrestraind passions, that the Reformation in England was so summarily achieved. The Pope once so proudly bearded by the monarch, the prestige of the Romish clergy was forever gone, and the people, except during one short interval, were ever after freed from their dominion.

This is not an occasion upon which to discuss the motives by which the unscrupulous monarch was actuated in the course he adopted, or how far a regard for religion held a place in his views. That he was well versed in the conflicting tenets and opinions, which at the time formed the subjects of so fierce a controversy between the votaries of the two religions, he has left behind him the most indubitable proof. Few could handle the polemical bludgeon, more adroitly, than the man who styled himself "Defender of the Faith." In that respect, indeed, it may be doubted if any one of the many upon whom that title has descended, could have boasted of equal fitness. Be that as it may, Henry may be said to have achieved a bloodless victory, and laid the foundation of an order of things, upon which, without any doubt, the present proud pre-eminence of England has been chiefly superstructed.

A very different aspect of things presents itself when we turn our eyes upon Scotland, at the time the reformed doctrines began to insinuate themselves into the minds of the few, who, from general enlightenment, were qualified to give them a rational and consciencious reception. With the throne based upon Popery, and in league, generally, with the most powerful Roman Catholic kingdom in Europe; with a nobility unlettered, proud, fierce, and wholly given to war—their religion possessing no more than a superstitions influence over their conduct; with a population existing in a state of serfdom, pliant to the will of their chiefs, but ignorant or regardless of every other obligation, it would appear upon a cursory view, that, of all the countries of Europe, the doctrines, precepts, and practice of pure religion would have found least favour in Scotland: yet it was not so. In no other country was the Reformation hailed with more hearty acclaim. It sped throughout the land with a celerity and effect, bearing a resemblance to that of primitive Christianity under the ministry of the Apostles and their immediate successors. The main cause of this success, will, it is believed, be found, under providence, in the peculiarities of the national mind. More capable of deep, close, and consecutive thought, particularly when it is applied to the solution of abstract propositions, than perhaps any other people of modern times—and alive, at the same time, to the loftier impulses emanating from whatever is true and purifying in morals, or beautiful and attractive in the regions of fancy and imagination—with zeal and courage to defend to extremity whatever principles their judgment and conscience may adopt—and bound together, in social and domestic life, by ties both warm and enduring, it may not be difficult to conceive what would be the effect upon a people possessed of such elements, when the Bible, with all its beauties and sublimities; its awful denunciations and soothing entreaties; its prophetic mysteries and enchanting narratives, was first unfolded to their newly awakened feelings and perceptions. England might have received the great boon with equal gratitude, but it was reserved for Scotland to accept it with a glow of enthusiasm, the written records of which form a beautiful page in the moral and religious statistics of our country.

Another cause, under Providence, of the rapid success of the Scottish Reformation—which, indeed, had its origin in the same source as that above stated—was the zeal and ability of its first teachers. This brief sketch admits not of even a simple enumeration of those intrepid men, by whom the ancient faith was uprooted, and the new one implanted in its stead.—The name of Knox, however, will occur to the recollection of every reader. This extraordinary man seems to have been raised expressly for the work in which he bore so prominent a part. Every faculty of his mind bore the characteristic traits of a Reformer. Highly intellectual, and initiated into all the branches of knowledge then taught in the schools, of a fearless and inflexible temperment, which sustained him unappalled, even when adversely confronted by a monarch on the throne, his course was onward and successful. He has been well termed the Apostle of the Scottish Reformation. The time, the *hour* had come, and the *man* stood forth; and his countrymen hailed him as the leader of the greatest moral revolution which Scotland has ever experienced.

By Knox and his coadjutors, the work of the Reformation in Scotland, as far as doctrine, formula and discipline were concerned, was carried up to the very point at which we find it at the present day. Certain things may have fallen into temporary abeyance, but not an item has been lost sight of by one section or another of the community. Every deviation has, from time to time, been struggled against, and it might not, perhaps, be too much to add, with no small amount of success. The cause, however, had a long and adverse course to run, ere it was permitted to settle down into peace. Nor was it from the remains of Romanism that its sufferings were derived—for Popery had ceased to give much disturbance in the land at the time of James' accession to the throne.

Although James was nursed in the very lap of Presbytery; yet it soon became apparent that that was not the religion of his choice. Nor was his preference of Episcopacy a thing to excite wonder. The Stewarts had all along entertained high notions of the nature and extent of their prerogative; and hence it must, from the first, have been almost certain that an arbitrary prince could never become sincerely attached to a form of church government so glaringly founded upon the purest principles of republicanism. James indeed professed Presbytery for a time, but he gave it no more than a hypocritical observance—and even that much only for a short time. As soon as he imagined his seat upon the throne to be firmly established, and when he became morally certain of succeeding Elizabeth on the throne of England, he boldly threw aside the thin veil under which he had endeavoured to conceal his long-cherished purpose, and rudely set himself to saddle his people with a form of religion which they considered to be only one short step in advance of Popery.

James has descended to us under the character of a somewhat imbecile person. An attentive perusal of his history, however, does not seem to warrant the conclusion; and a late writer of discernment has controverted it with much seeming success. He was without personal dignity, and had even no objections, upon what he thought a fitting occasion, to enact the part of a buffoon. The birch of Buchanan, however, seconded by an eager appetite on his own part, had, between them, furnished him with a fund of learning, and even knowledge, exceeded only by a few of his subjects, and probably far beyond the mark of his cotemporary monarchs. He had a considerable portion of the genius which seems to have been common to all the Stewarts; but he was totally deficient in that chivalrous courage by which the greater number of them was so eminently distinguished. The reports of foreign ambassadors to their respective courts, all agree in giving highly favourable opinions of his ability, both as a man and as a monarch. He was neither a tyrant nor a bigot, like his son Charles, but a good-natured, conversible man, fond of hunting, cockileikie, sheepshead and Scotch haggis, and by way of dessert, an occasional monkey-trick. His greatest fault was one common to most kings, even at the present day, an overweening sense of his own power. He would be an arbitrary monarch, and as he considered the republican spirit of Presbytery inimical to his claims, he became the persecutor of the religion of his native subjects. It was from this absence of severity in the disposition of James, that the early religious contests between him and his subjects, frequently assumed a character little short of farce,—himself the chief actor in the scenes. In a very different spirit was the contest carried on during the respective reigns of his son and grandsons—the savagery of which has not been surpassed in modern times

Few who may peruse the following papers, need to be reminded of the nature, and intent, of the solemn Covenants, entered into by the people of Scotland, during the reigns of James and his son Charles the first. The first of these engagements was simply a guarding against the machinations of popery, framed in the minority of James; but which having served the ends of its institution, was permitted, by the bulk of the nation, to be laid aside, among other half forgotten things; until the conduct of Charles the first, brought it again forward, new-modelled and extended, as a defensive weapon against the inroads of Episcopacy. The third edition of this famous instrument under the title of "The Solemn League and Covenant," so called to distinguish it from its immediate precursor, "The National Covenant," was called forth with the common consent of England and Scotland alike, when the undisguised tyranny of Charles united both countries in a league for their mutual safety. The success of Cromwell, however, and the Independents, very soon cut the link which connected the two countries, through this sympathetic engagement.

Perhaps there never was, either in ancient or modern times, an instance of a more throughly organised band than that exhibited by the people of Scotland, at the resuscitation of their former engagement, under the name of "The National Covenant," in 1638.—When this patriotic document, the composition of the wisest heads in Scotland, was laid before the people in Edinburgh, Glasgow, and the other large towns, such was their eagerness to adhibit their signatures, and such the earnest and enthusiastic feeling manifested by all classes, of both sexes, that it seems matter of astonishment, that any goverment should have ventured so to trample upon the principles and consciences of a nation, so firmly and universally united in the maintenance of a cause, in which they believed their interests, for time and eternity, to be indissolubly bound up. After receiving the signatures of many of the highest and noblest in the land, as also those of the middle and lower classes, men, women, and children, in the larger towns, copies of the covenant were carried into every district and corner of the land, and few, indeed, and insignificant, were those who did not literally rush forward to identify themselves with the grand national demonstration.— The scene of its subscription in the Greyfriar's church of Edinburgh, is thus described by a living historian. "When it had taken the round of the whole church, it was handed out to the immense multitude which had collected in the church-yard, and there being received with no less rapture than in the church, it was laid upon one of the flat monuments, so thickly scattered around, and subscribed by all who could get near it. It is said by one of the contemporary chroniclers, to have been a most impressive sight when the covenant was read to this vast crowd, to see thousands of faces and hands at once held up to heaven in token of assent; while devout aspirations burst from every lip, and tears of joy distilled from every eye." The divine right of enslaving men's consciences, was perhaps never more impudently asserted; nor have we any instance of its being met with a bolder or more uncompromising front."

That devoted generation could not have been called rebels, in any sense of the word. To worship their maker, according to the dictates of their consciences, was all they asked. Their expression of inalienated loyalty to their native sovereign, was as sincere as it was urgent and reiterated. It went for nothing, and Scotland had to pass through a period of fifty years persecution. They purchased their sacred rights at a high price, but it fell with still more direful effect upon the bigotted dynasty who exacted it.

It was not until the humblest petitions and remonstrances had failed to mollify the obdurate Charles and his advisers, that the covenanters resolved upon vindicating their rights by force of arms—and that the War Committee, the operations of a section of which forms the subject of the following pages, was first organised—and both the king and his subjects set themselves openly to prepare for war. In the spring of 1639, the War Committee commenced levying an army over the whole kingdom. In the words of the historian above quoted, "The chief covenanting nobleman in each county was placed at the head of each corresponding regiment, with the title of crowner; and the principal gentry were appointed to act as inferior officers."

Old compaigners, some of whom had served under Gustavus Adolphus, were engaged to drill and discipline the troops, and the most distinguished of these, Alexander Leslie, was appointed commander in chief; with private instructions, from the War Committee, not to assume a too independent authority over the proud nobility, lest their jealousy should frustrate that unity of action, so necessary in all military operations. Money, to the amount of two hundred thousand merks, was borrowed; their friends on the continent furnished them with arms for thirty thousand men, and a foundry for cannon was established in the suburbs of Edinburgh. The means, by which money was raised, will be sufficiently explained in this volume. Should these means appear, as they probably will, to have been inconsistent with the liberty of the subject, an excuse must be looked for in the nesessity of the case,—in other words in the all-absorbing importance of the object at stake.

It forms no part of the editor's design to enter upon a detailed account of the operations of this most eventful period of our national history. As above stated, his chief purpose was to connect the detailed operations of the Stewartry War Committee, with the general history of the period of its institution.— He is not aware that any document of the same kind has ever before been made public—and it gives him encreased confidence in the opinion which he himself entertained, that many of his friends, better qualified to judge than himself, were also of opinion that the merits of these sheets fully entitled them to a separate publication.

Unless the editor very much deceives himself, these papers will excite more than a single interest in the estimation of his readers. In the first place, they will give a nearer, and more distinct, view of the means and the machinery, by which our covenanted forefathers were enabled, successfully, for a time to withstand the vindictive efforts of their implacable and unprincipled sovereign; backed as he was by the crafty hierarchy, as well as by the haughty aristocracy of England;—and in the second place they will possess no common local interest.

There is no page in Scotland's history, since the days of Wallace and Bruce, calculated to excite a more glowing sensation in the bosom of Scotsmen, than that which records the events of this period.— War, on the part of the resistants, is here exemplified under the only colours that can ever justify it. It was a war in which every social, civil, and religious priviledge was at stake; and it was the noble stand made by our forefathers, at that day, which more than any thing else, perhaps more than all other things beside, has left us in the proud position we at present occupy—free, within the pale of just laws, to exercise our political and social rights—and free to worship our maker on our own mountain, without any to make us afraid.

It is true that many, perhaps most, among the rich and powerful of the land deserted the standard under which they had at first so boldly enrolled themselves; and it is also true that, up to this day, there have been among ourselves, those who scrupled not to justify that desertion; and in whose nostrils the word covenanter is an unsavory vocable. The different views which men have ever taken on the same subjects, is the mildest excuse that could be adduced in defence of the renegade covenanters of that day. They might have come to think that the matters contended for were too inconsiderable to balance what was contained in the opposite scale. To contend to the death for a principle is an undertaking fitted for only a few. While willing to admit of an excuse for men placed under such trying circumstances, let us turn with a double gratitude to the small, the heroic band who *never gave in.* That they have been branded as bigots and fanatics goes for nothing.—Supposing that there were some foundation in truth for such a charge, let us consider what it was, and who they were, who drove them to such extremities. They styled themselves "the Righteous," "God's Saints," and by other such like designations;—scriptural terms, by which we find the persecuted, in scriptural times, always designated; and because they did so, they have been called selfrighteous, bigoted, fanatical and intolerant. Let it be remembered that the Bible was their only book; and that from it they drew the main consolations of which their condition was susceptible. Its language was that upon which their own more serious or excited phraseology was constructed; and is it at all surprising that they should have freely applied it to their own forlorn condition: besides, if there were such a thing, at that time, as saints in Scotland, where else than among this persecuted remnant, were they more likely to be found? At any rate they were human, and being human, could we have expected that such treatment as they experienced, through a long series of years, was to leave them possessed of no other spirit than that of "saints triumphant"? To the utmost of their power, they gave blow for blow, and all brave men do so, when contending for life and its inalienable rights. The following language has been put into the mouth of one of their number, and quaintly strong as it may appear, it gives no more than a truthful account of the sufferings which thousands were at that period destined to undergo.

"Could I only record what I have suffered in myself and seen suffered by others engaged in the same cause, *that* record might prove a valuable heirloom to future generations. There is now a lull. The Laggs, the Dalyells, the Lauderdales, the Claverses, the Sharpes, and many others that I could name, have ceased to afflict God's heritage. Some of them have gone to their account, and all await it. It is not for me to judge them nor shall I. They have caused the blood of the righteous to flow as if it had been worthless as ditch water. *That* is the witness against them—and it is enough. That it has ascended from earth and claims retribution, who shall gainsay? That it has reached the ear of Divine justice, can there be a doubt? Earth has done with them, however; let us leave the settlement of the account to the judge of us all.—Whether the rest which at present we enjoy has been purchased at no more than its value, men will decide according to their lights. Peace is ever desirable—but there is a peace which is not peace. I have wrestled in prayer, I have yearned in my soul, nay, I have fought and shed human blood with carnal weapons. What I have seen and enacted might make a boarded book;—for many years of my long life were passed amid danger and distress. I have wandered over the moors and the mountains, derning in the caves, or cowering under the clifts and craigs of my native home-land; blinded by the lightening; deafened by the thunder; battered by the rains; tossed about by the winds; and gulphed in a snow wreath; and only escaping from these perils to be hunted by the minions of our godless rulers. It is over, for a time at least,—may it be forever! and I am left, at the age of three score and seventeen, (a rare wonder,) to think and speak of what is past. * * * * * It has been by the people of Scotland and not by the gentry, that the Kirk of Scotland, under providence, has ever been upheld; and I, though a shortsighted mortal, will venture to predict, that, should a time come, when the great bulk of the people shall have withdrawn their countenance from a law founded kirk, the prop of the upperfolk will be found of little avail in preserving the unction of its usefulness.—A state-kirk must ever be a stale-kirk. When men have once delivered the management of their souls' concerns into the hands of their earthly rulers, and submitted to be ministered to by whomsoever those rulers may appoint—without say or salvo in the matter—then farewell to manhood, farewell to religion—their place is in the Indies, where they say men's shoulders are bored, and where men are made to supply the place of ousen."

Over and above the merely historical interest which this little volume is calculated to awaken, the editor promises upon its being equally fitted to excite other items, of a more personal, and therefore of a more immediately pleasing nature. Many will be able to recognise the footprints of a remote ancestry, in situations, and under circumstances, of which they were previously ignorant; and will thus derive a fund of fresh associations—not only with the events and occurrences of a long bygone day, but also with the scenery of their respective localities, which from long familiarity had sunk to the level of mere commonplaces. To have to say, "Here, on this very spot where I now stand, centuries ago, stood one whose blood I know, or beleive, to be circling in my own veins. Here he acted so and so, under feelings which I can at this moment transfer from his breast to my own. His eye has rested on the same object upon which mine now rests; and has wandered over the surrounding expanse, under all the varied moods of mind which that exciting period would necessarily call forth; and which I can yet sympathise with and appreciate. In yonder village he must certainly have signed the Covenant's "National and Solemn League," perhaps with that blood; (many did so,) a portion of which is the source of my own vitality. And it must have been under that time-scathed tree that his dwelling once stood—where he lived, worshipped, suffered, (perhaps sinned,) and died, for a cause which if I think lightly of, I am not worthy of such a sire." To have to say something of this kind, and many may know themselves entitled so to say, must give double zest to the loco-historic gleanings, which the editor has been so kindly encouraged to lay before the eye of the public.

Many readers will also have their curiosity gratified by remarking the changes which have taken place in the occupancy and proprietory of the districts to which these papers so frequently refer. They will find the names of those apparently high in influence, whose evident descendants are now treading in the humbler walks of life; and even potent families which have ceased to exist in the localities over which they once bore high sway; or if we may venture to guess at the existence of a single relic of them, we find that relic sunk into the plain tradesman or humble artizan.—Such are the terms upon which human greatness holds its fickle tenure; an unceasing round of change; fortune's wheel at every revolution tossing from it at a tangent some proud mortal, who had vainly dreamed that he had secured to himself a permanent perch on its highest spoke. He will also find, and every good mind will be pleased at the recognition, other families or races, who have breasted and successfully surmounted the conflicting elements which whelmed even their superiors, of that day, in irretrievable ruin; and who are yet honourably engaged in supporting the framework of a very much improved social and political fabric.

It is scarcely possible, in this sketch, wholly to pass

over unnoticed the fortunes of one family, at the period in question certainly the most pre-eminent, in territorial influence within the bounds of Eastern Galloway: the editor alludes to the noble house of Kirkcudbright. Wide as their dominions then were, it is a fact that only one individual of the name, is now in possession of a single acre of their original territory. The title has merged in a highly accomplished lady, who believes herself to be the last representative of her far-descended ancestry. How far she may be correct in that conclusion may admit of question. Of a race once so numerous as to consist of thirteen branches, all acknowledged to have sprung from the root of Bomby, it is not easy to conceive how an heir to the title should not exist somewhere—unless the very name, the patronymic itself, should have ceased to be found within the boundaries. The editor is aware of one claim having been openly made within the last twenty years, in the person of the late Rev. John M'Clellan, minister of Kelton, and a native of the town of Kirkcudbright. That gentleman firmly and consciencіously believed himself to be the righteous heir to the title; and was understood to be in course of substantiating his claim, up to the period of his death in 1840. How far he had succeeded or fallen short, in his genealogical enquiries, is not within the editor's knowledge; nor is it now, in his opinion, of much consequence. Mr M'Clellan's death was speedily followed by the death of his only brother, and also of that brother's only son. If there are others who think themselves possessed of a claim, they have not, as far as the editor knows, openly avowed it—so that the collateral branches of this unfortunate family may now be considered the same as if they were wholly extinct.

KIRKCUDBRIGHT,
29TH DECEMBER, 1854.

MINUTE BOOK

KEPT BY THE

WAR COMMITTEE OF THE COVENANTERS

IN THE

STEWARTRY OF KIRKCUDBRIGHT.

MINUTE BOOK

OF THE

WAR COMMITTEE.

The Committie of the Stewartrie appoynted within the Stewartrie of Kirkcudbryt, halden at Cullenoch,[1] the 27th Junij, 1640. Larg[2] chosen preses.

Act anent the provisioune of the troupe horss.

The quhilk day the Committie ordaines, that, the troupe horss to be leviat furth of the Stewartrie for the service of the publict. That, the worste horss be worthe jc. lib. monie,[3] and whar horss are better to be apprysed be the Committie according to thair worthe, and the siller peyit thairon, aither in monie or security. And ordaines, that ilk trouper have for

1 Until a recent date the village of Laurieston went under the name of Clauchan-pluck or Cullenoch, and from its being situated nearly in the centre of the Stewartry, most of the Presbytery, and many of the County meetings were formerly held at this place.

2 Sir Patrick M'Kie of Larg, see Appendix.

3 All the money mentioned in this book is Scotch, except when otherways specified. It may not be amiss to state, that the proportion which Scotch money bears to Sterling is as one to twelve; consequently, one shilling Scots is equal in value to one penny Sterling; and one pound Scots to one shilling and eight pence Sterling. A mark Scots is 13⅓d sterling.— When xx, c or m are inserted after any number or sum, they express scores, hundreds or thousands—thus vj^{m} viij^{c} iiij^{xx} is six thousand, eight hundred and four score; ij^{m} ix^{c} iij^{xx} is two thousand, nine hundred and three score.

the twa pairt of the 40 dayes lone appoyntit be the Committie of Estaites xviij libs., conforme to the generall order; and that ilk horsman have for arms, at the leist, ane steill cape and sworde, ane paire of pistolles, and ane lance, and for fornishing thairof, ordaines to be given xx rex dollares.

Act anent the divisione of the said troupe horss amangst the parochess.

Ordaines, that, the equal halff of the 80 troupe horss imposit upon the shire and Stewartrie, be the generall instructiouns, be delyverit as followes, viz,,—xxiiij thairof above the water of Urr, and vj benethe the said water; and, that, the samyn horss be put furth at the sight of the persones efter-specifit; ilk ane within the severall boundes as is efter declarit, vix.,—furth of the parochess of

Carsfairne[1] and Dalry,	iiij.	Kenmure[2] & Erlistoune.[3]
Kelles and Balmaclellan,	ij.	Kenmure and Erlistoune.
Balmaghie and Partoun,	ij.	Balmaghie,[4] Bargaltoun,[5] and Mocherome.

1 See Appendix.

2 John, 3rd Viscount Kenmure, see Appendix.

3 Alexander Gordon of Erliston, see Appendix

4 John M'Ghie of Balmaghie. This family, from whom the parish derives its name, and who retained possession of extensive estates in its neighbourhood, till the end of the last century, were the descendants of an Irish chieftain, who, at an early date, settled in Galloway. When the church of Balmaghie, with all its lands and revenues, was granted to the Bishop of Edinburgh, the M'Ghies of Balmaghie retained their right to the patronage of it, under a charter granted by James VI in 1606; and about the year 1641, one of the family was minister of the parish.

5 William Grierson of Bargalton was Commissioner for the Stewartry of

Crocemichael, . .	j.	Kirkconnell.
Tongland Twenom, and Kirkchryst, . .	ij.	Kirkcudbryt[1] & Kirkconnell.
Buittle, Gelstoune, Keltoun and Kirkcormock,	iij.	Logan, Netherthird, Thos. Hutton of Arkland, and David Cannan of Knoks.
Kirkcudbryt, . .	j.	Lord Kirkcudbryt.
Dundrennan, . .	ij.	Lord Kirkcudbryt & Collin.
Borgue and Girthtoun,	iij.	Lord Kirkcudbryt, Kellie[2] and Carletoun.[3]
Anweth, . . .	j.	Cardynes.[4]
Kirkmabreck, . .	j.	Carsluthe & Cassincarrie.
Monegoff, . . .	ij.	Lord Galloway & Larg.

It is ordainit that these persones forsaid, snot onlie sie the forsaid horss, men, and mantainance put furth, each from thair divisone above mentionit, but also sie that they be put furth with armes and mantainance, conform to the former order.

Act for mantainance of the foote Sogers.

Ordaines, that, ilk of the sogers quhilk are to go furth to service, furth of the divisonies, for mantainance and his commanders, have the said twa pairt of the 40 days lone (to be peyit be the tennants and

Kirkcudbright in the Scotch Parliaments of 1644 and 1649. According to tradition, the Griers and Griersons of Galloway are descended from a branch of the clan Gregor.

1 Robert M'Clellan, 1st Lord Kirkcudbright, see Appendix.

2 John Lennox of Callie. The family of Lennox acquired the Callie and other lands in the Stewartry, by the marriage of Donald Lennox of Ballcorragh, to Elizabeth Stewart of Callie. Their second son William was infeft in the lands of Callie in 1469.

3 John Fullarton of Carleton, see Appendix.

4 John Gordon of Cardoness, see Appendix.

yeomanes, ilk ane for thair awne pairtes, according to their occupations, proportionablie upon the yeirlie rent to be laid on, the soume of nyne punds viijs. viijd.

Act in favores of Cardyness.

Ordaines Cardyness to cause clipe and intromit with some scheipe parteining to Bakbie, ante-covenanter, and to be comptable thairfoire to the publict, and that they remaine upon the ground whair they are until they be gotten sauld.

Act in favores of Netheryet.

Ordaines Netheryet to stop the cuting the woode of Cars: as also, the transporting of the woode and bark thairof, quhilk pertaines to the pretendit bischope of Edinburgh,[1] and to summons Johne Gowne, called keiper thairof, befoire the Committie the next Committie day, quhilk will be the sexth of July.

1 In the General Assembly, held at Glasgow in 1638, an act was passed declaring Episcopacy to have been abjured by the Confession of Faith of 1580, and that all titles, except those of Pastor, Doctor, Elder, and Deacon, with the offices depending thereupon, ought to be rejected and disallowed in the Church. At this Assembly, sentence of deposition and excommunication was passed against Mr John Spottiswood, pretended Archbishop of St Andrews; Mr Patrick Lindsay, pretended Archbishop of Glasgow; Mr David Lindsay, pretended Bishop of Edinburgh; Mr Thomas Sidserfe, pretended Bishop of Galloway; Mr John Maxwell, pretended Bishop of Rosse; Mr Walter Whytefourd, pretended Bishop of Brechen; Mr Adam Ballantyne, pretended Bishop of Aberdeen; Mr James Wedderburn, pretended Bishop of Dumblane; Mr James Guthry, pretended Bishop of Murray; Mr John Grahame, pretended Bishop of Orknay; Mr James Fairlie, pretended Bishop of Lismoir; Mr Neil Cambell, pretended Bishop of Isles; Mr Alexander Lindsay, pretended Bishop of Dunkell; and Mr John Abernethie, pretended Bishop of Cathness. At the passing of this sentence a sermon was preached before the General Assembly by their moderator, Mr Alexander Henderson, which sermon, together with the act of deposition, was published in 1762, in a small pamphlet entitled "*The Bishop's Doom.*"

The Committie of the Stewartrie foirsaid, halden at the said Cullenoch, the sexth day of Julij, 1640. Erlistoune preses.

Act contra Borge and Johne Gordoun, Ruscou.

The said day the Committie expelles the resounes preponit be Borge and Johne Gordoun, in Ruscou, and ordaines, that, they both goe upon service in the publict, as twa of the Captains that are to go furth of the Stewartrie, or else the Collonell, my Lord Kirkcudbryt, supplie Borge's place.

Act contra Cardyness.

The Committie also ordaines, that, Gordoun of Cardyness is to goe furth upon service in the publict as ane uther ane of the Captains.

Act James Gordoun.

Also ordaines that James Gordoun, brother german to Alexander Gordoun of Erlistoune, goe upon service in the publict, as the fourth Captaine.

Act anent the divisioune of the Sogers amangst the Captains.

It is ordainit that the Sogers, to be leviat furth of the Stewartrie be devidit amangst the Captains as is efter specifit, viz.,—the Collonell to have the parochess of Borge, Twenome, Kirkchriste, toun and paroche of Kirkcudbryt, out of the paroche of Tungland, iij men, and of these that are to be leviat under the water of Urr, fourtie; quhilk makes up his Lordship's company: Captaine Gordoun of Cardyness to have the parochess of Monegoff, Kirkmabreck, Anweth, Girthetoun and out of Tungland fowr: Captaine Johne Gordoun in Ruscou, to have Rerick, Buitell, Keltoune, Gelstoune, and Kirkcormock, and Crocemichael: and Captaine James Gordoun to have the parochess of Carsfarne, Dalry, Balmaclellan, Kelles, Partoun, and Balmaghie, and furth of the paroche of Tungland, one: and ilk ane of the saids thrie Captains to have of these that are to be leviat under the water of Urr, fourtie, quhilk will complete thair numberes.

Act anent the time and place of the rendevouez.

Ordaines, that, the rendevouez both of foote and horss be at the Milnetoun in Urr paroche, upon the sexteine day of this instant.

Act anent the payment of both foote and horss.

Ordaines the mantaniance of both foote and horss to be upon the rent proportionallie, viz.,—furth of ane thousand merks of rent, above the water of Urr, to pey for mantainance of ane horss fiftie merks monie, and for mantainance of the foote fifteen merks monie, quhilk is to be collected by ane commissioner in ilk paroche, and to be brought by the said commissionersto the rendevouez, time and place appoyntit, thair to be delyverit with horss and foote sogers to thame that shall have power to ressaive the samyn.

Act in fabours of Sir Patrick M'Kie.

Ordaines Sir Patrick M'Kie of Larg, knight, to uplift furth of ilk thousand merks of rent within the parochen of Monegaff and parochen of Kirkmabreck, and that for the fornishing horss, and mantainance of the samyn, within these paroches; and the person that shall happen not to pey willingly, according to thair proportion, these presents gives power to the said Sir Patrick to poynd thame thairfoir.

Act Lieutenant Colonell Stewart.

Approves the letter sent in favours of Lieutenant Collonell Stewart and his officers frae the Committie of Estaites,[1] for peyment to thame of the twa months provisiones, viz,—December and January, as containit in the letter, and ordaines the collectors to pey the samyn upon thair acquittance thereof.

Act for summoning of alleged monied men.

In obedience of the generall instructiones anent the tryall of what money may be had to borrow upon suiretie, ordaines, that, the persones underwritten be summoned in against the next Committie day,— Thomas Conchie, Johne M'Cauchie in Monegaff paroche, Robert M'Culloch there and Thomas Leitch there: Johne M'Dowall in Kirkmabreck: Johne Bell in Anwith, Johne Gourlay there: Johne Aires in Girthtoun, Andro M'Mollan there, Andro Carsane

1 See Appendix.

there: In Kirkcudbryt, Baillie Ewart,[1] Johne Fullartoun, proveist, Johne Carsane,[2] baillie, William Osborne, Agnes Pauline, Robert Kirk: In Buitlle, Cullengnaw, Wm. Gill, James Bonnan: In Dalry, Johne Newall, Johne Gordoun: In Kirkpatrick, Overbar, Lochinkit, Thomas Neilson, Auchenskeoch, and Netherbar: In Troquier, Johne Broune, elder and younger, Nicol Thomsoun and James Rig: In Lochrutton, Patrick Clerk, Johne Uchtarsone, James Clerk, Johne Pot, and Johne Aitken: In New Abbey, Johne Greggans, elder and younger, Johne Rig, Johne Stewart, younger.

---

1 See Appendix.

2 John Carson of Serwick, was repeatedly Commissioner for the Burgh of Kirkcudbright, both in the Scottish Parliaments and General Assemblies.

The family of Corsan, or Carson, have it handed down from age to age, that the first of their ancestors, in Scotland, was an Italian Gentleman of the Corsini family, who came into this realm with an Abbot of Newabbey, or Dulce Cor. in Galloway, about the end of the 13th century. Some versions of the family traditions say, that he was architect or master mason, at the building of the Abbeys of Holywood, and Dulce Cor, or Sweetheart, and also of the Franciscan Convent, and the old stone bridge over the Nith, at Dumfries—which were founded and endowed by Dervigild, daughter of Allan, Lord of Galloway, and mother of John Baliol, King of Scotland.

Among many other instances that might be given of this ancient family of Corsanes, appearing from authentic vouchers, this is one, Sir Alexander Corsane is witness to a charter granted by Archibald, called the Grim or austere Earl of Douglas, to Sir John Steuart, laird of Cryton of the lands of Callie; though the charter is without date, yet it must necessarily have been before the year 1400, when the granter of that charter died."

The Committie of the Stewartrie foirsaid, halden at the foirsaid Cullenoch callit Clauch-anepluk, xiij Julij, 1640 Erlistoune chosen preses.

Act for the Minister of the raigement.

Ordaines Bargaltoun and Collin to supplicate the Presbytrie that aither Mr Johne M'Clellane[1] or Mr Samuell Bell may be appoynted by thame as minister to the raigement.

Act anent the nomination of Commissioners in everie Paroch.

Seing that by the want of Commissioners in parochess the publict does smart. Thairfoire, thinks it neidful and necessar that thair be Commissioners, ane or mae, chosen in ilk paroche within the Stewartrie, wha shall have power within thair boundes to uplift the sogers, both the foote and horss, mantainance and armes, and to produce thame at the rendevouez, time and place appoynted; and, in caice any of the sogers alreadie nominate, or to be nominate, absent thamselffes at the uplifting, the said Commissioners shall have power, by thir presents, to upmak the number whar men may best be wantit. And ordaines the said Commissioners to plunder any persone that shall

1 See Appendix.

happen no to mak thankfull peyment of the sogers pey both for horss and foote; and that the parochinares assit the Commissioners for doeing thairof. The Commissioners undermentionit are appoyntit in ilk paroche, viz:—

Carsfarne, . .	Knockgray.[1]
Dalry, . .	Dabtoun, Waterside, Gairlarge[2] and Knockreock.
Balmaclellan, . .	Shirmers and Holm.
Parton, . .	Lagan, Cogourt, and Moch-erome.
Crocemichael, . .	Kirkconnell.
Balmaghie, . .	The Laird of Balmaghie.
Buittle, . .	Logan, Netherthird & Knocks.
Kelton, Gelston, and Kirkcormock,	Netherthird and Thomas Huttone.

1 "Alexander Gordon of Knockgray, a rare christian in his time. His chief, the Laird of Lochinvar, put him out of his land mostly for his religion; yet being restored by that man's son, Lord Viscount Kenmure, he told me the Lord had blessed him so as he had ten thousand sheep"—LIVINGSTONE'S MEMORABLE CHARACTERISTICS.

2 Alexander Gordon of Garlarge, and his two brothers Robert and John Gordon of Knoxbrax, are all said by Livingstone to have been worthy and experienced christians Robert attended the Assembly of 1638, as Commissioner for New Galloway, which burgh he afterwards represented in several Parliaments and public meetings. He is supposed to have been the worthy and religious gentleman mentioned in Kenmure's Heavenly Speeches, as having visited Lord Kenmure a few days previous to his death.

After the defeat of the Covenanters at Rullion Green in 1666, Robert and John Gordon, the sons of the Laird of Knockbrax, who had been there taken prisoners along with Major John M'Culloch of Barholm and others, were sentenced to be executed, and their heads were ordered to be sent to Kirkcudbright and exposed on the principal gate of that Burgh. The right arms of all the prisoners were to be sent to Lanark, where the Covenant had been taken with uplifted hands, and affixed to the public ports of that town. It is said that the two brothers, when thrown of the ladder, clasped each other in their arms and thus expired. The joint testimony of those who suffered along with them, may be found in Naphtali, and a full account of the trial is contained in Samson's Riddle.

Rerwick, . .	Lord Kirkcudbryt, Colin and Newlaw.
Kirkcudbryt, . .	Haircleugh[1] and Gribdae.
Borge, . . .	Carletoun and Robertoun.[2]
Girthetoun, . .	Kellie and Ruscow.
Kirkmabreck, . .	Carsluth[3] and Cassincarrie.[4]
Moneygoff, . .	Sir Patrick M'Kie, and Machermoir.
Tungland, . .	Bargaltoun and Barcaple.[5]
Suddick and Colven, .	Fairgirth and Milnbank.

1 Telfair of Haircleugh. This family was long in possession of lands in the parish of Kirkcudbright. According to a tack contained in the Register of Deeds of the Stewartry, dated 4th March, 1585, the two merk and a half and the fourtie penny lands of Drumore, in the parish of Dunrod, were set in lease by John Telfair of Haircleugh to his son William Telfair.

2 William Gordon of Roberton, with his brother-in law John Gordon of Largemore, joined the Covenanters in the ill advised rising at Pentland, and was killed in the combat. John was severely wounded and, having lain for some nights after the engagement in the fields, he died a few days after he reached home, from the hardships he had then suffered.

3 The Browns of Carsluth, were one of the most ancient families in Galloway. Over the armorial bearings above the door of Carsluth, is the date 1364 and under them that of 1581. Gilbert Brown, the last Abbot of Sweetheart Abbey, was descended from this family. About the year 1600 he was engaged in a controversy with the celebrated John Welsh, (the son-in law of John Knox,) who was sometime minister at Kirkcudbright and afterwards of Ayr. Brown was an inflexible Catholic, and having been apprehended about 1607, was sent to Blackness, and, a few days afterwards, transported to Edinburgh Castle. Shortly after that he was permitted to leave the Kingdom, and died in France in 1612.

4 From an entry in the Burgh Records of Kirkcudbright it appears that Janet Brown, in June 1647, granted a lease of remunciation, in favour of her husband Richard Muir of Cassincarrie of her life rent of the lands of Corss and Spittle, and Ferry Boat of Cree, with the parts, penticles and privileges thereof. The Muirs were proprietors of the lands of Cassincarrie, at a very early date. There is a copy of a Wadset in the Register of Deeds of the Stewartry of Kirkcudbright, dated 7 May, 1586, in which the twenty shilling lands of Mackwilliam, in the parish of Kirkmabreck, are wadset by John Muir of Cassincarrie, and Janet Muir his spouse, to John Halliday in Glen of Skyreburne, for the sum of six hundred merks. The said lands to be held *frie blanch* of them, they paying yearly the feu maill of the said lands to the Abbot of Dundrennan and his successors.

5 After the restoration of Episcopacy, David Arnot of Barcaple was repeatedly fined for nonconformity. His brother, Mr Samuel Arnot, who

Act contra the entrant Tenant.

Being debated whether the entrant tenants, at Whitsounday last, or the tenants that removed at the said term, shall be lyable for mantainance of the foote sogers, findes, that the entrant tenants shall be lyable and not the removear.

Deposition of the allegit monied men.

The persones underwritten being put upon thair oathe, gif they have any monie to lend, upon suiretie, to the use of the publict, depones as follows :—They are to say :—

Johne Greggane, eldir, in Newabbay, hes onlie j^c merks monie of the realm.

Johne Greggane, younger, hes onlie about five punds sterling monie.

Johne Broune, eldir, at Brigend of Dumfries, about xj^xx. merks.

Nicoll Thomson, younger, iiij^c. lviij lib. xiijs. iiijd. and hes to pey furth thairof to his creditors, about j^c merks.

Johne Broune, younger, thair, hes iij^c. merks, which he is owand to creditors.

---

was minister of Tongland, was one of those who were then turned out of their parishes, and he, together with Mr John Wilkie of Twynholm, afterwards held many field preachings or conventicles in the presbytery of Kirkcudbright. Both of these ministers attended the great communion at Irongray. Mr Arnot was one of those who escaped from the defeat at Pentland, and, along with many other ministers and gentlemen belonging to Galloway, was declared outlawed by a proclamation dated 4th December, 1666. Both Mr Arnot and his brother afterwards fled to Ireland, in order to escape the persecution of the times.

Ordaines the minister of Newabbay to tak Johne Briges oathe, what monie he may spare.
Johne Bell of Arckland, nihil.
Johne M'Douall, nihil.
Johne Gourlay, xx lbs.
Johne Bell in Clauchried, xl lib.
Thomas Conchie, iij$^{xx.}$ lib. xs.
Johne M'Knacht hes xxxvij lib.
Johne Cutlar[1] in Dundrennan, nihil.
Johne Cutlar, younger, hes xxj lib.
Thomas Telfeir[2] in Heselfield, ij rex dollars.
Robert Kirk, xx merks.
Johne Newall of Barskeoch, xl lib.
Johne Gordoun of Barr, ix^{xx} merks.
Mr Train,[3] v^{c} merks.
Patrick Clerk, nihil.

---

1 The family of Cutlar is of very old standing in the parish of Rerwick. According to tradition, the first of the family who came to that parish, was employed in sharpening the tools of the masons engaged in the erection of Dundrennan Abbey, and that he thereby acquired the name of Cutlar. William Cutlar, of Auchnabanie, is mentioned in the Register of Deeds in 1587.

2 Mr Alexander Telfair, minister of Rerwick, wrote a curious pamphlet, entitled "A true relation of an Apparition, expressions and actings of a Spirit, which infested the house of Andrew M'Kie, in Ringcroft of Stocking, in the parish of Rerwick, in 1695." The truth of the various circumstances of this relation, is attested by a number of the ministers of the presbytery and others. Mr Telfair was most probably related to the Telfairs in Hessilfield. In 1687 he was engaged as chaplain by Sir Thomas Kirkpatrick of Closeburn, but shortly after proposing to go to England, he came and stopped in the neighbourhood of Auchencairn, where he exhorted and preached to the Inhabitants. The Curate of Rerwick at this time was greatly disliked by his parishioners, and they, upon receiving news of the revolution, rose in a body, and having ordered him to leave the manse within twenty four hours, immediately put Mr Telfair in possession of it, and choose him as their minister.

3 Mr Alexander Train, minister of Lochrutton, attended the Assembly of 1638.

Johne Clerk, nihil.

James Thomson, in Kirkandrews, vij^{xx} merks, of which he is owand to his Maister xlviij merks.

Thomas Neilsone[1] of Knockwalloch iiij^{c} merks.

William M'Mollan in Netherbar, v^{c} merks.

Cuthbert M'Mollan of Drumness, iiij rex dollars.

Johne Aires, nihil.

James Gordoun and Johne Hamiltoun absent.

---

1 Mr Nesbit, in his Heraldry, states, that, according to "common tradition, three brothers of the surname of Oneal, came from Ireland to Scotland, in the reign of Robert the Bruce, where they got lands for their valour, and their issue changed their name a little, from Oneal to Neilson; for Oneal and M'Neil are the same with Neilson. For the antiquity of this family, I have seen a precept granted by James Lindsey of Forgirth, to infeft John Neilson and his wife Isabel Gordon, in the lands of Corsack in Galloway, in the year 1439. Also a charter of confirmation of the lands of Corsack of the date 20th of July, 1444, by Sir John Forrester of Corstorphin, to Fergus Neilson, son and heir to John Neilson of Corsack."

John Neilson of Corsack, being warmly attached to the presbyterian worship, and not attending the ministrations of Mr Daglish, the curate of the parish, was subjected to severe oppression on account of his nonconformity. In 1666 he joined the Covenanters and accompanied them to Pentland, where he was taken prisoner and carried to Edinburgh. On the 4th of December, he was taken before the Privy Council, and they entertaining the idea that the rising at Pentland was an organized conspiracy, ordered him to be immediately put to the torture of the *boot*, in order to extort a confession. Crookshanks, in his History of the Church, says, "Corsack was dreadfully tormented, so that his shrieks would have melted the hearts of any except those present, who were so far from being moved, that they still called for the other touch" Six days after he had been thus barbarously maltreated Mr Neilson was brougth to trial and sentenced to be hanged at Edinburgh on the 14th of December. Sir James Turner, whose life he had been instrumental in preserving, attempted to procure a mitigation of his sentence, but all his endeavours were rendered fruitless by the curate of the parish, who represented Mr Neilson as being the very ringleader of the disaffected in Galloway.

After Mr Neilson's execution, his wife and family suffered great oppression. Woodrow says, "His lady being in Edinburgh after her husband's death, Maxwell of Milton came to the house of Corsack, with thirty men, and took away everything that was portable, and destroyed the rest, and turned the family, and a nurse with a sucking child, into the fields." The family continued to be exposed to a continuance of distresses by soldiers being quartered upon them, sometimes as many as thirty horsemen at a time, and by fines imposed upon them for not attending the ministrations of the Curate.

Letter frae the Estaites,

Giving warrand for the third pairt of the fourtie dayes lone to be peyit furth of the tenth penny, quhilk is of the date, the 10th Julij instant, as follows.

Ryght Honorabill, We ressaivit your letter, and having considerit the desyre thairof, and for satisfying of your desyres thairin mentioned, we are content that ye pey to your own raigement, whilk is to come furth under the commandement of my Lord Kirkcudbryt, that third pairt of the pey quhilk is appoynted to be peyit to thame in monie, and that as weill to the officers and commanders as to the common sogers, and that ye keipe the general commissar frie of that burding and pey the said raigement, sae lang as your tenth penny lasts, being always comptable to the collector generall for the samyn. This is the easiest and best cowrse we could think upon for your ease and satisfactioun. Ye shall pey, furth of your tenth penny, the third pairt as weill of the horss as foote, and what remains frie, we expect, assuredlie, that ye will use all possible diligence to send the samyn heire. Your affectionate friends,

(sic subscribitur,) Burley.
Naper.
S. Murray.
Forbes of Leslie.
Jas. Sword.

The Committie of the Stewartrie foirsaid, halden at Mylnetoune of Urr,[1] xviij Julij, 1640. Collonell preses.

Act in favours of Mr Gavine Hamiltoun, Minister at Kirkgunzeon.

Gives power to Mr Gavine Hamiltoun, minister at Kirkgunzeon, to buy sae much guids frae my lady Herries[2] as will pey the monie due for my ladyes landes, and that both for fornishing of the horss and foote.

Act contra Lady Kenmure.

Ordaines, that my lady Kenmure's dewties, grassounes and uthers, in the paroche of Tungland, shall pey the same as any uther's rentes or lands in the paroche, and that for the mantainance of both horss and foote.

1 The Milnton of Urr, at this time a considerable village, was a Burgh of barony and had its market cross. As it lay nearer the eastern boundary of the Stewartry than any other village, of any magnitude, it was genarally used as the place of rendevous for all forces which were intended to be sent either to England or the Borders.

2 Elisabeth, eldest daughter of Sir Robert Gordon of Lochinvar, sister of the first Viscount Kenmure, was married to John, eight Lord Herries, who, joining Montrose, was excommunicated by the General Assembly in 1644. Balfour in his Annales of Scotland, states that he was declared forfeited, by the Estaites, on the 11th February, 1645, and on the 8th December of the same year "The Estaites ordaines Commissarey Leuingstone to guie to the Ladey Harries for this zeir, in respecte of her necessities, 2000 merkes."

The Commitie of the Stewartrie foirsaid, halden at Drumfries, the xxiiij day of Julij, 1640. Collonell preses.

Act anent the Threive.

Ordaines ane letter to be sent to the generall, what course shall be taken with the house of the Threive.[1]

Act in favours of Mr David Ramsay, Minister at Newabbay.

Ordaines ane warrant to be given to Mr David Ramsay, minister at Newabbay, to Johne Stewart of Schambellie, to mak peyment, to the said Mr David of his ordinar stipend of viij$^{c.}$ and fiftie merks monie, and that for the yeir of God 1639.

Act in favours of Johne Stewart.

Anent the supplication presented by the said Johne Stewart of Schambellie. Efter due consideration thairof, ordaines that he retein in his own hand his factor's fie, dew to him by the pretendit Bischope of Edinburgh,[2] and that for the crop and yeir of God 1638; and, thairfoir, uplift from the rents that perteined to the said pretendit Bischope.

1 See Appendix.

2 Charles I., in 1633, having founded the Episcopal Bishopric of Edinburgh, prevailed upon Sir Robert Spottiswood and Sir John Hay, to whom the Abbey of Newabbey, with its revenues, had previously been

Act in fabours of Lochearthure.

Ordaines, that, Lochearthure and his wyff be not plunderit, by any persones whatsomever, in respect of the band given by thair cautioners of the dait heirof.

Act anent the parochess under Urr.

Ordaines, that, the parochinares under the water of Urr, nominate the minister and twa of the maist considerable men in each paroch, to come to the Committie upon Tuesday next, to give thair valuatiouns under thair hands, with certification to thame that compeirs not the said day, they shall never be held thairefter to give in thair valuatioun, and the Committie to impose upon thame, as they think fittest, bothe for horss and foote.[1]

granted by James VI., to resign their pretensions to it, and gave a grant to that See of the Abbey of New Abbey, together with the Churches of Buittle Crossmichael, Kirkpatrick Durham Urr, Newabbey and Balmaghie.

Sir Robert Spottiswood was the second son of the celebrated Arch-bishop of that name. In 1622 he was appointed a Lord of Session, under the title of Lord Newabbey, in the room of his father, who resigned in his favour; and, upon the death of Sir James Skene in 1633, was elected president of that court. Upon the triumph of the Covenanters, he fled into England, and on the apprehension of the Earl of Lanark, then Secretary of State, at Oxford in December, 1643, he received the seals of office from the king and acted as Secretary. Having joined Montrose at Athol in 1645, he was taken prisoner at the battle of Philliphaugh, and beheaded at the market cross of St. Andrews, on the 20th January, 1646.

1 It would appear that the ancient Valuation Rolls of the Stewartry of Kirkcudbright and the Shire of Wigtown, were drawn up about this time. In an act of the Estates of Parliament, dated the 20th June, 1650, the monthly maintenance of the forces then on foot, for the month of July, is to be paid, in the respective shires, according to the valuations approven by the parliaments, except, only, the Stewartry of Kirkcudbright and the Shire of Wigtown, which are to pay according to the Valuations lately given in by them.

The Committie foirsaid, halden at the place above specifit, the xxv day of the said month. Collonell preses.

Act against those that alleges thamselffes to be overvaluit.

Ordaines, that, notwithstanding of the grevainces given in by those that alleges thamselffes to be over-valuit, that the collector of the levie ressaive conform to the said first valuatioun, without any deductioun, and that within the haile of the Stewartrie.

Act for Commissioners to meit with those of Nithisdaill and Annandaill.

Makes choyse of the Lairde of Balmaghie, Bargaltoun, and Cassincarrie, to be Commissioners for the Stewartrie, to meet with the Commissioners of the Sheriffdomes of Nithisdaill and Annandaill, anent the publict affaires.

Act anent the Inbringing of Monie.

The said day the instructiouns efter specifit are sent to the haile paroches within the Stewartrie, to be proclaimit upon Sunday first, with the names thairof that are ressaivers, the tyme and place appoyntit.

1. At Edinburgh the xv day of Julij 1640 yeires, these of the Committie of Estaites have appoyntit and ordainit, that the act of parliament, that letters of horncing poynding and captioun be executed against

collectores, valueares and uthers quha doeth not thair deutie or mak peyment of thair tenth pairt, in manner efter specifit, viz.,—

2. Against the collectores, for not making compt, rekoning and peyment of that quhilk they have ressavit and giveth not in the roll and names of these quha have not peyit.

3. Against the valueares, for not valueing conform to the said act of parliament, quhilk is, aither upon the heritor's oathes, or upon heritor's declarations, under thair handes; with certification, that they quhilk dallie shall be confiscatit; and for not-delyverie of the said valuations.

4. Against the not-peyers of the said tenth pairt, by apprehending thair persones, poynding thair own proper gudes, or poynding thair grund.

5. And because, efter all ordinar meanes are usit to making peyment quhilk is deu, yet, such is the unwillingness and delaying of some, to the evill example of uthers, that, monie cometh not in to serve the present time; thairfoir, it is thought fit, for mantainance of the army presentlie on foote, for preservation of the religione and libertie of the Kingdome, that all these quha have any monies, shall lend the samyn for the publict use, in manner efter specifit, viz.,—those within the Sheriffdomes of Edinburgh, Haddingtoun and Linlithgow, within four dayes efter intimatione; those of the Sheriffdomes of Aire, Stirling, Lenerk, Rainethrou, [Renfrew] Fyfe, Arguile, Berwik, Perth, Roxburgh, Peibles and

Selkirk within sex dayes efter intimatione made thair; and sicklyke those of the Sheriffdomes of Aberdeine, Banff, Murray, and Inverness within ten dayes efter intimatione. Such monie as shall be pledged shall be frie of any common burden by detentione of any pairt of the annual rent, but, shall have thair full annual rent frie of any burden or detentione.

6. Secondlie, they shall have full annual rent frae the lending thairof, as the samyn shall be ressaivit within the said spacess requirit as afoirsaid, to the term of Whitsounday next to come, as for a haile yeir, notwithstanding ane good pairt is past.

7. Thirdlie, they shall have such securities as they shall pleise, desyre, or crave thamselffes, so that what persones they shall crave to binde, for whatsomever they lend, shall give their personall bandes for the samyn, and these persones, quha shall binde to thame, shall have the haile presbytrie or shire bound for thair releiff, and the presbytrie or shire shall have the Estaites bound to releive or repey thame.

8. And sicklyke, gif it shall be provit that any have monies and will not lend the samyn, it is ordainit that the act of parliament be put in execution against thame, especiallie in that poynte, that all these quha can be tryit to have monie and will not lend the samyn, as foirsaid, the dilatores and findares out are to have the ane half and the uther half to be confiscate for the publict.

9. And sicklyke, it is appoyntit, that all the silver worke and gold worke in Scotland, as weill to burgh

as landwart, as weill noblemen, barrones and burgess, as uthers, of whatsomever degrie or qualitie they be, be given in to the Committie at Edinburgh, or thame they shall appoynt to ressaive the samyn, upon such securitie for repeyment as the said Committie and they shall aggrie, at the prycess following; and for this effect the Committie of War within each Sheriffdome, and the Magistrates within each burgh, with concurrance of the Ministrie, quha must exhort and give warning out of the pulpites to the parochinares, are appoyntit to call befoire thame any such persones as hath any silver or gold worke, that inventar may be maid of the weght and spacess thairof, and securitie given for the samyn, with declaration alwayes.

Lykeas, it is heirby declarit, that these quha hes any silver or gold worke quhich they crave raither to keip for thair ane use than delyver the samyn to be coinzed, shall have power to redeime the samyn at the prycess efter following, viz:—fiftie sex schillinges for the unce of Scotts silver worke, fiftie aught schillinges for everie unce of Inglis silver worke, and xxxiiij lib. vjs. viijd. for everie unce of gold; the samyn being aither producit befoire the Committie at Edinburgh, or befoire the Committie of War in each Sheriffdome, or befoire the Magistrates of everie burgh, and inventar maid thairof, or else declarit be the pairtes, under thair handes, and monie presentlie peyit at the rates and prycess foirsaid, for the quhilk monie securitie shall be given for repeyment thairof, and that frie of any burden, as said is.

And, in case anie hath doubill gilt worke, and curious wrought worke, and cannot get monie to redeime thame, it is heirby declarit, that, the said gilt and curious worke being deliverit to the said Committie, shall not be melted nor disposit upon, befoire the term of Whitsounday next, betwixt and quhilk time the owner thairof shall have power to redeime the samyn, at the prycess foirsaid, peying alwayes the annual rent thairof, sae lang as the samyn shall be unredemit.

And the said silver and gold worke to be all given in, aither to the Committie of Estaites, or Committie of War within each Sheriffdome or Presbytrie, or to the Magistrates of each burgh, within eight days efter intimatione be maid thairof, aither at the severall merkat crocess, or by touk of drume,[1] or by advertis-

---

1 At this time all royal burghs, at the election of their office bearers, appointed a town drummer, the duties of whose office combined those of the modern bellman, and a burgh officer. He also beat the drum through the town at certain stated hours. The following extract from the Burgh Records of Kirkcudbright, dated October 1600, is the first notice which occurs of a drummer being appointed in that burgh, previous to that time all proclamations were made by the town piper

"The quhilk day, Alexander Corkirk is chosin and seit drummar for ane zeir, for the quhilk, he sall haif ten libs. of fie, and his meit throu the toun, and that thai that hes not housses pey him iijs. iiijd. ilk day for his meit, and gif thai refuis, ordaines the refussar to be poyndit for vjs. viijd. thairfoir."

"The quhilk day Fergus Neilsone is seit toun pypar for ane zeir, his dewtie usit and wont. [x libs,] provyding he and the drummar pairt the Zule wages [Christmas' boxes] betwixt thame"

The following extract from the same records may perhaps be not uninteresting. as it shows how much the pride of the Magistrates was wounded by thair drummer having presumed to engage himself with another party.

"Burrow Court, xj Maij, 1642, be the Proveist and Baillies.

"The qlk day the Proveist, Baillies and Counsellors being convenit, and heiring ane report going, that Robert Rig, drummar, had agriet with Capitane Wm. M'Clellane, to have gone with him to Ireland, to be his drummar to ane fit company, and thairefter did agrie with the tounship to

mentes frae the ministeres out of the pulpites, with certification to those that shall not give in, nor redeime the said silver wark or gilt wark, within the said space, the samyn shall be confiscatit for the publict use

Instructiones anent the Borrowing of Monie, and the Ressaving of Silver Plait,

Conforme to the act of the Committie, the 23 *Julij*, 1640.

1. It is appoyntit, that in everie burgh, the Magistrates within the samyn, desyre some sub-

be thair drummar for ane yeir. Quhairupon, they have severall tymes convenit the said Robert Rig befoire thame, and having requyrit him to declair the vieritie, he ever hithertill conceillit and denyit the samyn untill this present day; and being overagain requyrit to declair the veritie, he confessit that his father did first persuade him, and sterit him up to tak on and agrie with the said Capitane M'Clellan, to be his drummar to his companie, and for that effect to goe over with him to Ireland; and, thairefter, his said father, sterit him up, and persuaded him to conceill the first agriement and tak on of new again with the said tounship, to be thair drummar for ane yeir. Quhairin the Magistrates and Counsell findes that baith the father and the sone have dealt verie treacherouslie and unhonestlie with the toun, and ought to be examplarie punishit, to the terror of uthers, to comit the lyke in tyme cuming; and, thairfoir, they decern and ordaine the said Robert Rig, to be presentlie put in the stockes in the mercat place, and remaine thairin untill the setting of the sune; and, thairefter, to be brocht back again to the tollbuithe, and keipit thair in cloce ward, untill the tounes farder plessor. And because Stephane Rig, his father, did counsell, entyse, persuade, and steir up his sone to such unhonest dealings, thairfoir, they decern and ordaine the said Stephane Rig, to remove himselff, his wyfe and sone out of this burgh, and libertie heirof, betwixt and the term of Whitsounday, now instantlie approaching, under the pain of such farder censure and bodilie punishment as the said Magistrates and Counsell shall think maist fit and expedient to be inflicted upon thame, or anie of thame, that shall at anie tyme thairefter be fund within the said burgh and libertie thairof. Quhairupon the Judges ordanit act.

W. Glendonyng, Proveist.
Johne Ewart, Baillie.
J. Carsane, Baillie.
George Callendar, Counseller.
William Halliday, Counseller.
C. Meik.

stanteious, honest men, quha shall sit at leist foure houres everie day, as they shall desyre, in the toll-buithe or some open chamber, to ressaive any lent monie, or silver or gold worke quhilk shall be de-lyverit to thame.

2. The saids Magistrates shall cause give secur-itie, to the persones frae whom silver or gold worke is ressaivit, at iij lib. the unce for the Scots silver worke, and iij lib. ijs. for the unce of Inglis silver worke, and xxxviij lib. for the unce of gold, quhilk shall be given in to be melted, and not redemit conforme to the former act, peyable at Mertinmas, and interest thairefter, frie of all common burden, and that, till the tyme ane sufficient securitie be given thame by the Committie of Estaites, quhilk shall be dune immediatlie efter the report of the said Magis-trates tickets, upon deliverie of the worke or monie.

3. These quha lend monie may desyre the per-sones whom they crave to be bound to thame, quha shall grant band, and be releivit according to the act.

4. Some must be appoyntit in everie Presbytrie, by the Committie thairof, quha must sit foure houres everie day, for doeing the lyke, anent lent monie, silver or gold worke, and securitie given in manner foirsaid and the releiff to be as said is.

5. Intimatione must be maid by open procla-matione, touk of drume, or out of pulpites, of the persones, dayes and places appoyntit for attendance, and giving satisfactione, as said is; quhich persones shall sit down within xxiiij houres efter the intima-

tione, and are to sit for the space of ten dayes, within quhilk space everie one, within the said severall presbytries and burghs, shall be oblischit to come and give in thair monie and silver plait, with certificatione containit in the foirsaid act, quhich is confiscation.

6. The said persones, so appoyntit, must send in weiklie all the monie and plait they ressaive, to the Committie at Edinburgh, upon the publict charges, to be peyit by the collectores at the deliverie thairof or thair diligence; that it may be knawn whair the fault is and the samyn remeidit in tyme.

And these presents to be publishit by the Magistrates of everie burgh, and by these of the Committie of everie Sheriffdome or Presbytrie, or by the Ministeres or Reiders efter sermone.

Letter anent Non-Covenanters, and Sitting upon Civil Affaires.

Dated 12*th August*, 1640.

The said day the Committie ordaines the letter underwritten to be registerit in the buikes of the Committie, quhairof the tenor followes:—

Ryght Honourabill.—We ressavit yours. As for answer to the first pairt,—whair ante-covenanters offeres now to joyne in the publict, being fried of farder censure,—pleid your minde, and advyse it not that any of that kynd sould be so slichtit ower, quhilk were ane verie unequal cowrse, that these that hes been still refractorie, and the greatest instruments of of all the troubill of the countrie, and quha hes saved all the charges, and been at ease frae the beginning of the business, and now quhen they are necessitat and hes nae uther refuge, offeres to contribute; that we cannot admit of; but thair rents and estaites must be exposit upon to the use of the publict.

As for the next,—anent those quha are frustrate of thair just debtes,—thair is no remedie for that but to use legall executione be horneing, poynding, apprysing, or utherwayes perseuing the debtores befoire the Sheriff or Judge ordinar; and treulie, it were incumbent to you of the Committie of everie divisione, in respect to the generall calamatie throw want of justice, to advert particularlie that justice be administrate, and necessar and trew debtes satisfied, and gif your ordinar judges be deficient, being desyrit be you to doe justice, it is your pairt in caicess of necessitie, to bring the pairties befoire you and sic order and credit keipit within your boundes, sae far as you are able.

And for the last,—concerning the want of Commissionar Deputes,—you must have ane warrand for that effect frae the Commissonar generall, quhilk shall be authorizit frae the table, and gif ye cannot get ane warrand frae him befoire he goe into Ingland, upon advertisement, ye shall have orderes frae the table to that effect, to the maist dilligent and weghtiest amangst you for that charge, quhom pleiss to nominate, quha must give assurance for dilligence, compt and reckoning. In the meantyme, until warrand be gotten, ye will doe weill to tak notice of ante-covenanters' rentes, and put your collectores and some uthers to worke for putting of the samyn to the best avail.

Thus we have answerit the haile poyntes of your letter as far as we can, and rests your affectionate freindes,

(sic subscribitur,) Loure.
Craighall.
Edward Edgar.
James Scot.

Act—William Glendonyng.

Having considerit the benefit that the countrie, and especiallie the toun of Kirkcudbryt[1] may enjoy by some judicious persone to be chosen as Captaine within the said toun, for commanding of thame; with the consent of the said toun, the Magistrates thairof doeth elect William Glendonyng[2] as Captaine within the said toun, for doeing of all charges appertaining to be done be ane Captaine within the said toun, in all tyme heirefter during thir present troubills.

1 For Extracts, from Burgh Records, connected with the Civil Wars, see Appendix.

2 The family of Glendonyng are of very old standing in the Stewartry of Kirkcudbright. William Glendonyng, provost of Kirkcudbright, attended the Assembly of 1638, as Commissioner for Kirkcudbright, which burgh he afterwards represented in several Parliaments and other public meetings. He was one of the Commissioners for the Kingdom of Scotland, who being in England in 1649, were ordered by the Scottish parliament to proceed to Holland to treat with Charles II. but were arrested by Cromwell at Gravesend when about to embark.—His elder brother, Robert Glendonyng, was Town Clerk of Kirkcudbright.

Their father, Mr Robert Glendonyng, would appear to have been closely connected with the family of Drumrash, as we find him in the Register of

Act—James Gordoun, Clerk depute.

The said day admitts and ressaives James Gordoun, notar, as clerk depute to the Committie, during the absence of Robert Gordoun, principal clerk, quha hes given his oathe, de fideli administratione.

Deeds in 1588, subscribing as witness in a Marriage contract between John Gordon appaient of Muirfad, in the parish of Kirkmabreck, and Jane Glendonyng, daughter of John Glendonyng of Drumrash. Mr Glendonyng succeeded John Welsh as minister of Kirkcudbright in 1602. The following copy of his engagement as minister in Kirkcudbright, is extracted from the Burgh Records, and dated the 14th July 1602.

"The qlk day the Proveist, Baillies and Counsell, all in ane voice, bindes and oblisches thame and thair successores to content and pey to Mr Robert Glendonyng, minister, zeirlie, during his lyfetyme and remaining minister at the paroche kirk, the soume of ane hundreth poundis monie of this realme, at twa tymes in the zeir, Whitsounday and Mertinmas, in twa equall portiouns; the first term being at Mertinmas next to come: and that as stipend deu to him, be the said burgh, for serving the cure of the ministerie at the said kirk of Kirkcudbryt, and for his manse gleib and kirklands thairof, qlk he shall not seik during the space foirsaid. Off the qlk soume, Androu Quhyteheid and his aires, shall pey zeirlie, to the treasurer of the said burgh, twentie merks monie at the terms foirsaid, equallie for the relieff of sae meikle of the said soume of j^{c} lib. Quhair. upon the said minister and tounship desyrit act."

In the life of Mr Welsh, it is said that "while he was in Kirkcudbright, he met with a young gallant in scarlet and silver lace, (the gentleman's name was Mr Robert Glendonyng,) new come home from his travels, and much surprised the young man by telling him he behoved to change his garb and way of life, and betake himself to the study of the Scriptures, which at that time was not his business, for he should be his successor in the ministry at Kirkcudbright, which, accordingly, came to pass sometime thereafter."

Kirkcudbryt, xxiiij August, 1640.

The said day Carletoun, Bargaltoun, and Knockbrax, Commissioners appoyntit be the Committie for ressaiving of silver worke, and borrowit monie, sat down in the tollbuithe, the tyme and place appoyntit, at ten houres, and sat until twa houres efter noone.

The said day none comperit, in manner and to the effect foirsaid.

Act—Johne Gordoun.

The said day, the Committie of the Stewartrie, anent the supplication presented by Johne Gordoun; apperrand of Erlistoun, shawing that, whairas, he, the last yeir bygane, being employit as ane of the Captaines within the Stewartrie, and that he and his officeres was upon service twa monthes tyme, during the quhilk space his officeres were peyit by himself, and also disbursed for his sogers ane certain soume conforme to his accompt, desyring that he may be peyit of the said disbursements, conforme to his said accompt producit with the said supplication. The quhilk supplication being seine and considerit by us, we doe ordaine William Griersone of Bargaltoun, collector of the tenth and twentie pennie rent, to pey to the said Johne Gordoun, the said accompt, upon his acquittance thairof, subscribit with his hand.—Subscribit by ane great many of the Committie.

The Committie foirsaid, halden at Mylnetoun in Urr, xxv August, 1640. Erlistoun preses.

Troupe Horss.

The said day, the said preses interrogating Dalskearthe, gif he would put out to the armie the troupe horss restand by that paroche, [Troqueer,] altogidder refuissit to doe the samyn.

Thomas Rome of Irongray, Commissioner for Irongray paroche, being requirit to put out the troupe horss restand by that paroche, refuissit to doe the samyn.

The paroche of Kirkbean wants ane troupe horss, j.
Colven and Suddik ane uther, . . j.
Lochrutton and Newabbay ane, . . j.

Citation of Commissioners.

The said day, the said preses, for the foirsaid neglectis, citates the said Dalskearthe, commissioner for Troqueer; the said Thomas Rome commissioner for Irongray; William Lindsaye commissioner for Colven and Suddick; Robert Maxwell of Cavence,[1]

1 John Maxwell, Bishop of Ross, was a son of the Laird of Cavens in Nithsdale. He studied at St. Andrews, and being bred to the church, first obtained the parish of Murthlack, and was afterwards appointed to be one of the ministers of Edinburgh. Having entered into the designs of Charles I. and Archbishop Laud with regard to the church government of Scotland, he was appointed a privy councillor, and promoted to the bishopric of Ross in 1633 On the 4th December that year he took his seat as an Extraordinary Lord in place of the Earl of Menteith. He was afterwards appointed a commissioner of the exchequer, and aimed, although unsuccessfully, at the

commissioner for Lochrutton; Johne Stewart of Schambellie, commissioner for Newabbay; Johne Chartres of Barnecleuche, commissioner for Terregles; to compeir befoire the Committie of Estaites, within ten dayes thairefter, to answer for thair neglectis.

---

office of treasurer, then in the possession of the Earl of Traquair. The Bishop of Ross was one of the chief promoters of the Service-Book and Scottish Liturgy, in the composition of which he had also a considerable share. During the tumults which ensued on the promulgation of that obnoxious ritual, although not actually maltreated, he was greatly alarmed, and repairing to London gave his advice there in favour of coercive measures. The king having been forced to permit the meeting of the General Assembly in October 1638, the Bishop was again sent to London on the part of his fellow prelates, to devise measures for their common safety; and he is supposed to have drawn up the declinator of the Assembly's authority, which was afterwards lodged on the part of the Bishops. This pleading, however, proved unavailing, and, like the rest of his brethren of the episcopal bench, the Bishop of Ross was, on the 10th December 1638, deposed and excommunicated, on the ground, 'that beside the breach of the caveats, he was a public reader of the liturgy in his house and cathedral; that he was a bower at the altar, a wearer of the cap and rochet, a deposer of godly ministers, an admitter of fornicators to the communion, a companion to papists, an usual player at cards on Sabbath, and once on communion day; that he had often given absolution to persons in distress, consecrated deacons, robbed his vassals of above 40,000 merks, kept fasts each Friday, journeyed ordinarily on Sabbath, and that he had been a chief decliner of the Assembly, and a prime instrument of all the troubles which befel both church and state.' He was, in the following year, declared an incendiary and an enemy to his country by the Estates, but was notwithstanding this promoted by Charles to the bishopric of Killala in Ireland on the 12th October, 1640. On the breaking out of the Irish rebellion in the following year, Bishop Maxwell was turned out of his house, plundered of his goods, and left by the rebels naked and wounded. The kindness of the Earl of Thomond, however, enabled him to reach Dublin, where he preached for some time. He joined Charles I. at Oxford in 1643. and, according to Baillie, was the King's ordinary preacher there. He had formerly attempted to support the royal cause by a pamphlet entituled, Sacro-sancta Regia Majestas, and this he now followed up by a violent attack upon presbytery in another pamphlet called Issachar's Burden. He was appointed to the episcopal see of Tuam on the 30th August 1645, but did not enjoy his preferment long, having been found dead in his study on the 14th February, 1646, a few hours after he had received intelligence of some disaster to the royal cause, grief for which was supposed to have occasioned his death. Burnett states him to have been a man 'very extraordinary, if an unmeasured ambition had not much defaced his other great abilities and excellent qualities.'"—Senators of the College of Justice.

Act in favours of the said Commissioners.

The Committie upon hoipes that the saids Commissioners will use greater dilligence, continues the citation foirsaid, until the next Committie day, ordaining thame all to be present then, utherwayes that this citation stand good, and that bye and attour the using of dilligence.

Act contra Johne Gordoun of Beoche.

Ordaines Johne Gordoun of Beoche, to underlye tryall, for the alledgit ryot committit by him upon George Levingstone in Quintenespie,[1] and to compeir the next Committie day; and ordaines Carstraman and his man to be citat as witnesses; and that Quintenespie citate the said George against the said day.

Citation of Hamiltoun and Gordoun.

Ordaines Glaisteres to citate Johne Hamiltoun of Auchenreoch, and James Gordoun of Lochinkit, to the next Committie day, for sic causes as shall be laid to thair charge.

Act—Captaine Johne Gordoun.

Empowers Captaine Johne Gordoun for redelyverie of the armes to the paroche of Buittle, alledgit to be ressaivit by him frae thame the last yeir.

1 The family of Livingston have long held lands in the parish of Balmaghie. In the Register of Deeds, Thomas Livingston is mentioned in 1588, as having borrowed a hundred pounds from John Glendonyng of Drumrash, for which he granted him security on his lands of Quintenespie.

Act—Erlistoun.

Doeth allow the desyre presentit by Erlistoun, anent the lending ij or iij^c. merks to ane destressit brother of the ministerie, be anie within the Stewartrie quha hes monie to lend.

At Kirkcudbryt, in the tollbuithe thairof, xxix August, 1640.

Johne Lennox of Kellie.

The quhilk day, deliverit to the said Commissioners, by Johne Lennox of Kellie, twa silver piecess, ane paire longe weires, nyne silver spoones, broken and haile, with ane stack of ane spoone, Scots worke, weghtan xxviij unce and iiij drops, whairof delyverit back of evill silver iij unce.

At the pairt fensed, the last day of August, 1640.

Knockbrax.

The said day, delyverit to the said Commissioners, by Robert Gordoun of Knockbrax, sex silver spoones, Scots worke, weghtan vj unce xij dropes.

At Kirkcudbryt, the first September, 1640. The Committie foirsaid, halden in the toll-buithe thairof. Kirkconnell preses.

Mocherome.

The said day, delyverit by George Glendonyng in Mocherome, to the said Commissoners, xj spoones, Scots worke, weghtan xiij unce iij dropes.

Act contra Beoche and George Levingstone.

The said day, Johne Gordoun of Beoche, and George Levingstone in Quintenespie, being baithe conveinit for committing ane ryot, and injuring ane anither, upon Fryday bypast, and Alexander Gordoun of Carstraman, and James M'Burnie being ressaivit as witnesses; being baith sworne.

Carstraman depones, that, George cryit—Wha is that that rides, Carstraman and unhonest Beoche—and that he [Carstraman] rode his way, and looking back saw thame striking at uthers with thair swordes; he came back, and when he came back, Beoche's sworde was broken and his finger bluiding, and he desyrit frae him his sworde, whilk he refusit to give.

George M'Burnie declaris, that, George cryit—Unhonest Beoche and adulterous Beoche—and that Beoche drew first his sworde

The Committie findes and ordaines, that, they be baithe committit to warde, presentlie, during thair plessor.

Act contra Grissell Gordoun.

Ordaines Grissell Gordoun, spouse to the umq[le.] minister of Urr, to present her silver worke, viz:—the twa piecess that was bought by the paroche of Urr for the use of the Kirk, and sex silver spoones pertaining to the aires of the said minister, and that notwithstanding of any reassones proponit in the contrair.

Act contra Marione M'Clellane.

Ordaines Marione M'Clellane, wyff of the late James Ramsay, to present her bairnes silver worke, and that notwithstanding of the reassones preponit in the contrair.

Act contra James Gordoun in Lochinkit and Johne Hamiltoun of Auchenreoch.

The quhilk day, James Gordoun in Lochinkit being convenit for conceilling of monie, in prejudice of the publict, and lending of the samyn to ane uther partie, viz, to George Rome; doeth alledge that the suirtie quhairupon the samyn was lent was subscribit and endit befoire he knew of anie sic order for lending of monie to the publict.

It was answerit, that, he was lawfullie commandit and desyrit to come to the Committie and lend monie upon suirtie to the publict, befoire the writting was passit.

Replyes, that, the minute of contract was foure or fyve dayes subscribit befoire his citation,

Ordaines James, the next Committie day, to produce the said minute of contract, with the principal contract passit betwixt him and the said George Rome.

And also, ordaines Johne Hamiltoune of Auchenreoch to doe the lyke, viz., to produce the writts passit betwixt him and the said George Rome.

Citation by Erlistoun of the Commissioners.

The quhilk day, Erlistoun, of new, again citates Collin, commissioner of Rerwick paroche; William Lindsay, commissioner for Colven and Suddick; Johne Chartres,[1] commissioner for Terregles; Johne Stewart, commissioner for Newabbay; Dalskearthe and Johne Broune, commissioners for Troqueer; Hew Maxwell in Torrorie, commissioner for Kirkbean, David Cannan, commissioner for Buittle; to compeir befoire the Committie of Estaites, at Edinburgh, the viij day of September instant; thair to answer for thair neglect for not out-putting of the troupe and baggage horss ilk ane of thame for thair awn pairtes.

1 The family of Charters claim to be descended from Thomas of Longoville, (the Red Rover) who was taken prisoner by Wallace, on his voyage from Kirkcudbright to France. On Wallace's return to Scotland, Longoville accompanied him, and attended him through all his campaigns. After Wallace had been betrayed into the hands of Edward, Longoville swore he never would depart from Scotland till he should be avenged upon the foes of Wallace. He afterwards joined Robert Bruce and was with him at the taking of St. Johnstone, where, being the second man that entered, he

" With charter'd lands was gifted by the king,
From whom the Charters ever since do spring."

In 1585 Robert Charters of Kelwood sold the lands of Dunrod in Senwick, to William M'Lellan in Balmangan, under letters of redemption, for the sum of 400 merks; and the same lands were let by William M'Lellan to Charters, until they should be redeemed, for the yearly payment of ten bolls six pecks of beer small measure.—REGISTER OF DEEDS.

The Committie of the Stewartrie foirsaid, halden ij September, at the place foirsaid. Kirkconnell preses.

Act contra Dalbeattie.[1]

Ordaines Dalbeattie to bring monie for the baggage horss of the paroche of Urr, the next Committie day.

Act contra David Cannan.

Ordaines David Cannan of Knocks, commissioner of Buittle, for detaining in his hove, one of the sogers that was appoyntit to have gone furth of that paroche, to pey to the collector of the Committie the soume of twentie punds, and to remain in ward until the samyn be peyit.

Citation of Johne Ewart.

The said day, Erlistoun citat Johne Ewart, baillie of Kirkcudbryt, and Johne Carsane, also baillie, to compeir befoire the Committie of Estaites, on the aught day of September instant, and thair to answer for thair neglect in not putting out thair troupe horss.

Valuatioun of Kirkcudbryt.

Soume of the Valuatioun of the Toun of Kirkcudbryt, iij$^{m.}$ iij$^{c.}$ libs.

---

1 A family of the name of Reddick were long proprietors of the estate of Dalbeattie. John Reddick of Dalbeattie is mentioned in the Register of Deeds in 1588, as one of the tutors of Gordon of Muirfad.

Act—Kirkconnell and uthers.

The Committie doeth empower Kirkconnell and the rest of the Captaines of the parochess, to borrow monie, frae any whatsomever, for fornishing of armes within thair boundis.

Act—Erlistoun.

Ordaines Troquhan, Shirmeres, Holme, Barscobe,[1] Crogo and the rest of the gentrie of Balmaclellan to meet with Erlistoun, as Captaine of that paroche, at all tymes requisite, and to contribute to him thair best assistance and advyse for furtherance of the publict in that paroche.

Erlistoun.

The said day delyverit by Erlistoun, to the Commissioners, ane silver peice and ane dussane spoones, weght, ij pund ij unce ix dropes. Maire, ten silver spoones, weght, ane pund ij unce ix dropes. Maire, ane silver coupe and ane silver peice, weght, xiij unce xj dropes.

Item.—Maire by Erlistoun, vj silver spoones, ane paire belt heides, ane pair silver weires, and foure uther little peices of silver, broken and haill, weght, xj unce xv dropes.

Item.—Maire by him, sex silver spoones, weght, x unce xiij dropes.

Whairof delyverit back again of evill silver, ane pund ane unce xiij dropes.

---

1 See Appendix.

Act contra Johne Gordoun of Beoche.

The said day, the said Committie, for the foirsaid ryot, committit by Beoche upon the said George Levingstone, and for the upbraiding of the table, by saying that he was committit to ward without ane fault; ordaines him to stay in ward until the next Committie day, and to pey fiftie marks of fyne; and also, ordaines that the said Johne stay in ward, untill the tyme that he finde sufficient cautioun, by band, that he shall absteine frae haunting and frequenting privatelie or publictlie, by nicht or by day, in any soirt of beheaviour with Margaret Levingstone, spouse to Johne Merteane, in all tyme heirefter, and that under the payne of ane thousand merks, to be peyit to the Committie, to the use of the publict, in case of failzie.

Act contra George Levingstone.

The said day, ordaines that the said George Levingstone shall stay in ward, until he find securitie to the Magistrates of Kirkcudbryt, that, upon the first mercat day he shall sit in the stockes, in tyme of mercat, betwixt ten and twelve houres befoire noon of the day, and that he shall upon the Sounday thairefter, stand in the gorgets at the kirk of Balmaghie, at the gathering of the congregation, and to pey to the Committie of fyne ten merks monie.

Act contra the Captaines of the Parochies.

Ordaines, that, the Captaines of the parochess against the next Committie day, bring in ane perfect roll subscribed by the ministrie and thamselffes, of all pretendit bischopes, ante-covenanters, and uther unfreindes rents, guides and geir, within thair bounds and ordaines, that, the Commissioners' commission shall be intimate by the ministeres of the parochess.

Carstraman.

The said day, delyverit to the Commissioners by Alexander Gordoun of Carstraman, xij silver spoones, weght, ane pund half ane unce. Maire, ane gilt coupe, Inglis worke, weght, v unce xiiij dropes.—Delyverit back ane unce and v dropes.

Kirkconnell.

Delyverit by Kirkconnell, ane silver coupe and cover, weght, ane pund iij unce. Maire, ane silver coupe and silver dish, weght, ane pund nyne unce iiij dropes, Inglis worke. Delyverit back again ane pund iij unce.

Dabtoun.

Delyverit by Dabtoun, vij spoones, weght, ix unce vj dropes.

Waterside.

Delyverit by Andro Chalmers of Watersyde, vj spoones, ix unce and ane halff.

Act contra Largmoire.

The said day, Roger Gordoun of Largmoire, for his contumacie in not coming to the Committie, being lawfullie citate thairto, baithe by the kirk officer and Erlistoun, is decernit in xx merks monie of fyne.

The Committie of the Stewartrie foirsaid, halden at Kirkcudbryt, the third day of September, 1640. Kirkconnell preses.

Act—Carletoun.

Ordaines the laird of Carletoun to tak these sogers within the paroche of Girthetoun that was nominatit to have gone furth of that paroche and Zedwick.

William Maxwell.

The said day William Maxwell, brother to Robert Maxwell of Culnachtrie, becomes in the Committie's will, for his lying sae lang out in not subscryveing of the covenant.

Act—Dalzell.

Ordaines Johne Dalzell in Staniedykes, to remain in ward until he find caution that he shall bring, against the next Committie day, Margaret Sampell's testification that he is her hired servant.

Act in favours of George Levingstone.

Continues the sentence against George Levingstone, until the next Committie day, and farder during thair plessor.

Act—Largmoire.

Mitigates Largmoire's fyne to ten merks monie, to be peyit presentlie, or else that he pey the foirsaid twentie merks monie.

Barnecleuche.

The said day, delyverit by Johne Charteres of Barnecleuche, sex silver spoones, Scots worke, weght, ten unce.

Knockbrax.

The said day, presented by Robert Gordoun of Knockbrax, to the said Commissioners, in name of William Gordoun of Robertoun, sex silver spoones, and uther work, weght, ix unce, ane drope.

Quintenespie.

The said day, delyverit by George Levingstone of Quintenespie, sex silver spoones, weght, ten unce.

Act—Mocherome.

The said day, decerns Alexander Gordoun of Carstraman, to content and pey to George Glendonyng in Mocherome, the soume of xxij lib. xiijs. iiijd. as the rest of the pryce of twa oxen bought by the late Hessilfield, frae the said George, about xiiij yeires since, or thereby; and ordaines the captaine of that paroche to sie the said George satisfied thairof.

Carletoun.

The said day, delyverit by Johne Fullarton of Carletoun, ane silver peice, Scots worke, ane gilt silver saltfat, with xiiij silver spoones, weght, twa punds nyne unce and ane half unce.

Bargaltoun.

Delyverit by William Griersone of Bargaltoun, xv silver spoones, weght, xxiij unce and ane drope.

Cardyness' Wyff.

Delyverit by the Lady Cardyness, in name of her husband, ane silver coupe, ane stak of ane fann, and sex silver spoones, weght, xv unce xv dropes. Delyverit back ane unce xiiij dropes.

Robert Gordoun.

Delyverit by Robert Gordoun, for himselff and certain uthers, certain silver worke, weght, thrie punds thrie unce xij dropes. Delyverit back v unce twa dropes.

The Committie of the Stewartrie foirsaid, halden at Kirkcudbryt, the tenth September, 1640. Collonell preses.

Act contra Precipitatores.

The said day, it is ordainit, that quhosoever, at anie tyme heirefter, precipitates and begins on purpose befoire ane uther be endit shall pey presentlie xijs.

Act—Commissioner.

Ordaines ane Commissioner to be sent to Edinburgh, to the Committie of Estaites, to present to thame certain grievances.

Act—Collonell.

Desyres the Collonell to write to the Committie of Nithisdaill, to intromit with the tenth and twentie penny rentes of the ten parochess under the water of Urr.

Erlistoun.

The said day, delyverit by Erlistoun, ane rental of non-covenanters' rentes within the paroche of Dalry.

George Levingstone.

George Levingstone presentit ane uther within the paroche of Balmaghie.

Act contra Laggane.

Ordaines Laggaue to present, the next Committie day, the contract passit betwixt him and the laird of Partoun, whairby he pretends right to the said Partoun's lands, with ane rentall of the dewties of the lands thairin contained, with ane inventar, and compt of his ressait and discharge of the dewties of the said lands, for all yeires bygane.

Schambellie.

The said day Schambellie peyit to Erlistoun, for the halfe of ane troupe horss furth of the paroche of Newabbay, quhilk was not put furth, ane hundred punds.

Act in favours of certain Captaines.

The said day the Committie, taking into thair consideratione that they had laid double charge upon certain persones, in respect that the maist pairt of the gentrie was thought to have gone upon service as volunteirs; doeth ordain that Shirmeres be Captaine within the paroche of Balmaclellan, and Erlistoun liberated of that charge during Shirmeres' abyde at hame. And the laird of Kellie, with the assistance of his sone, Captaine within the paroche of Girthetoun and Carletoun liberated of that charge. And that Kilquhennady be Captaine of Kirkpatrick-Durham, and Glaisteres liberated of that charge during Kilquhennady's abyde at hame.

Act contra Lochinkit.

The same day, comperit James Gordoun in Lochinkit, and produced the contract maid betwixt George Rome and him, of the date the fourth day of Julij, and confessit that the minute was not subscribit befoire his citation by Robert Gordoun, quhilk citation was upon the seventh day of the said month.

And also the said day, the said James, being convenit for conceilling of monie, in prejudice of the publict, as said is, efter that he was citat for lending thairof to the publict upon suirtie; confesses he has dune wrong in lending of his monie any uther way than to the use of the said common cause; and thairfoir he becomes actit in the Committie's will for his censure. The said James Gordoun has subscribit thir presents with his hand, day yeir and place foirsaid.

(sic subscribitur,) James Gordoun.

Act—Mr Hew Hendersone.

Ordaines, that, the house in St Johne's Clauchane, callit the Black Hall, be given to Mr Hew Hendersone,[1] minister of that paroche, efter the appryseing of the worthe thairof, by foure honest men, to the use of the publict.

The Committie of the Stewartrie foirsaid, halden at Cullenoch, the xxiiij day of September, 1640, by ane sufficient coram. The Collonell preses.

James M'Ghie.

The said day, James M'Ghie, tutor of Balmaghie, became actit in the Committie's will for not subscryveing of the generall band.

Act contra the Not-peyers of the Tenth Penny.

Ordaines, that, quhosoever pey not thair tenth penny betwixt and Wednesday first, shall pey ane third maire than the said tenth.

---

1 In 1643 a petition having been presented to the General Assembly, subscribed by a great number in the north of Ireland, setting forth the deplorable condition they were in through want of the ministry of the Gospel, and desiring that some ministers, especially such as had been driven away from them by the persecution of the prelates, might be sent to preach to them; the Assembly commissioned a number of ministers, among whom were Mr Hugh Henderson, minister at Dalry; Mr William Adair, minister at Air; Mr John M'Clellan, minister at Kirkcudbright; and Mr James

Act for Inbringing of the Testaments.

Ordaines, that, the Captaines of the parochess bring to the Commissioners ane note of the definoch testaments; and also, ordaines, that the said Captaines intimate the Commissioners' commission in thair paroche kirkes.

Letter frae the Estaites,

For converting of the tenth and twentieth penny of the parochess under Urr to the South Raigement, dated, 16*th September*, 1640.

Ryght Honourabill,—We have given orderes, and usit all the means we can, upon how that the south raigement may be mantainit, and has allowit the maist pairt of all that is due the publict furth of these pairtes for manteanment thairof, and yet, nevertheless, thair sogers are disbandoning for want of manteanment, quhilk is ane discouragement to the countrie, and ane encouragement to the enemie, and ane meanes to expose these pairtes to hazart and rewing.

Thair are ten kirkes of the presbytrie of Drumfries, quhilk lyes in Galloway, and within your lordship's divisione, the helpe quhairof would doe much good to the manteanment of that raigement, for the tenth and twentie penny of the said kirkes, with the rentes

Hamilton, minister at Dumfries, to repair to the north of Ireland, to visit, instruct, comfort and encourage the scattered flocks of Christ, and, if need be, to try and ordain such as should be found qualified for the ministry.—Mr Henderson was appointed minister to Col. Stuart's regiment in 1645, and was settled as minister in Dumfries in 1648.

of ante-covenanters, papists, and recusants thair, will be thairs; being duly and exactly uplifted. Quhair-foire we will heirby intreat your lordship, as ye respect the standing of that raigement, that ye will be pleasit, with all dilligence, to cause collect and intromit with the saids haile tenth and twentieth penny, with the rentes and gudes of all papists, ante-covenanters and rescuants in that boundes, and to apply the samyn for manteanment of the said raigement. And thir presents shall be to your lordship ane sufficient warrand; for what ye shall pey to the Commissioner of that raigement, upon his discharge, shall be allowit. Or, gif your lordship think it troublesome to you to meddle with the uplifting thairof, we desyre that your lordship, in that caice, will be pleasit to permit the said Commissioner, with the said Lieutenant Collonell's assistance, to uplift the samyn to the effect foirsaid.

So, being confident that your lordship will be carefull in the performance of ane of thir twa alternatives, we commit your lordship to God. Your lordship's affectionate freindes.

(sic subscribitur,) Argyll.
Wigtoun.
Marschell.
Cowpar.
Loure.
Balmirrinoe.
Jaffray.
Edward Edgar.

Act anent the Out-coming of Horss, as weill conforme to thair Rentes as Volunteires.

At Edinburgh, the last day of Junij, 1640 yeires. These of the Committie appoyntit by the Estaites of Parliament, taking into consideration how the instructiones for putting out of horss may be best effectit, conforme to the generall order, hath appoyntit and ordainit, that everie man, as weill to burgh as landward, shall send out ane sufficient and able horss and man, appointit with jack and lance, or with pistolles and carabine, and that according to two thousand merks of rent, conforme to the valuatiouns quhilk shall be the rewle of the mustares.

And sicklyke, it is appoyntit that everie heritor and tennant shall put furth thair best and maist fitting horss for that use. Naither shall any man whatsoever be sufferit to keip any good and able horss for the troupe at hame, but must aither send him out as ane of thair proportioun for thair rent, or utherwayes come presentlie out thairupon thamselffes, or some uther freindes as volunteires, or utherwayes sell thame at ane reasonabill rate for the use of the countrie.

And because, barrones and gentilmen of good soirt are the greatest and maist powerfull pairt of the kingdome, by quhas valure the kingdome hath ever been defendit, we do maist earnestlie requyre and expect, that everie barron and gentilman of good soirt shall come to the armie in thair own persones, or at leist send thair ablest sone, brother, or freind. And, that all noblemen, gentlemen and uthers, may be encouraged to come out as volunteires in sua good ane cause, for mantainance of religione and preservatioun of the libertie of this antient and never conqueirit kingdome, which we are all sworne to mantain; it is earnestlie desyrit that all brave cavaleires will tak the business to hart, and considder that now or never is the tyme to gaine honour and eternal reputatioun, and to saive or lose thair countrie.

Lykeas, it is heirby declairit, that quha shall so come out as volunteires, (having put out thair proportiounes in good and able horss, as said is,) shall have libertie to serve upon thair own best horss thamselffes, as volunteires, and shall have corne to thair horss out of the common magazine, and good quarteres for thamselffes, and shall not be put to toilsome dewtie.

So that it is heirby declarit that no good horss shall be sufferit to stay at hame, upon any kynd of pretext whatsoever, with certification that these wha shall fail in any of the premises shall not onlie be censurit as the Committie of Estaites shall appoynt, but will also be repute as loyterares or averse frae the good cause.

Act anent Maisterless Men, Beggares, Loyterares at hame, and uthers refuseing to goe out, being enrolled.

At Edinburgh, the last day of Junij, 1640 yeires. These of the Committie appoyntit by the Estaites of Parliament, taking into consideration the manifold abuses committed, and great prejudice ariseing to the whole kingdome, by some, quha, being nominate and enrolled to goe out maks shiftes, aither by running away, or lyeing obscure and hid by thair parentes, freindes or uthers.

And sicklyke, by some uthers quha have gone out and returned back, or runneth away frae thair colores.

And thirdlie, by those quha are idle and maisterless fellowes, naither at trade nor serving in the countrie, or under command as sogers.

For remeid whairof; it is ordanit and appoyntit, that the Captaines or Commissioners of ilk paroche, with concurrence of the heritors, whair any such shall loyter and refuise to goe out, as they are or shall be nominate and enrolled, be fund, shall cause apprehend thame, and bring thame bound to thair company; and quhaever shall be known to harbour, lodge and entertain thame, shall pey as followeth, viz:—everie heritor shall pey ten punds; everie yeoman and cottar five merks; to be exacted by the captaine or commissioner of the paroche, or by any the Committie of War in the shire or presbytrie shall appoynt.

It is lykewayes appoyntit, that all those quha shall run or steall away from thair collores, without ane pass, shall be apprehendit, whairever they can be fund, and brought to thair company, and punishit, conforme to the articles of militarie discipline.

And for remeid of the thirde; it is appoyntit, that no man travell in the countrie, or be harboured, or lodged in any place, as weill to burgh as landwart, quha are maisterless and not in service in the countrie, or under command of some Captaine; heirby declaring, that, it is not permittit to any idle man or vagabond, as weill beggares as uthers, to goe in the countrie, or get ressait, or harbour, or entertainment in any place, but that everie man tak himself to some calling, or to some maister or some captaine or Collonell. And gif any idle man, quha is able of bodie to worke or doe service, shall be fund, he shall be apprehendit by any officer, justice of the peace, commissioner of paroches, or any heritor within or without burgh, and put in jayll till he tak himself to some trade, maister or captaine. And for that effect, nae hostler, heritor or yeoman, within or without burgh, shall ressaive any such maisterless or idle man. The officers shall be fyned in fyve merks, toties quoties; the heritor shall pey twentie punds; and the yeoman or cottar shall pey foure merks, for ilk failzie. And that all these idle and maisterless men may be known, no man of that degrie shall be suffered to goe through the countrie, or be harboured in any place, within or without burgh, without ane testimonial or certificate, aither frae his officer, gif he be under command as ane soger, or by his maister, minister, or some man of note, quha are known to be heritores and freindes, or honest men, weill affected to the good cause.

And sicklyke, that the ministeres and elders in everie paroche, shall tak particular note of all the men within thair paroche, and gif any shall be fund aither idle, or quha have run away frae thair collores, the elderes shall be halden, under the payne of fyve punds, to give up his name to the minister, quha shall intimate the samyn upon the efternoon to the haill paroche, that the Commanderes, Commissioners of parochess, justices of the peace, or heritores may cause apprehend thame and put thame in jayll as said is.

And gif any of the said persones, quha, aither refuseth to goe out being namit and enrolled, or these quha runneth away frae thair collores, or idle or maistless men, shall be ressavit, harboured, or entertained by thair parentes freindes and kinsfolk, in that caice, the said ressettares and entertainers shall be fyned according to thair qualitie, at the discretione of the Committie of War, viz:—gif he be ane heritor or substantious soccarer or yeoman, ane hundred punds, as oft as they shall be warnit and desyrit and failzie in performance. And gif the said officers, justices of peace, commissioners of parochess, ministeres and elderes shall be negligent, they, or any of thame, for neglecting or overseing the said pairties, delinqueutes, shall be censured, viz,—the said ministeres and elderes by the presbytrie; and the said officers by the Counsell of War, and the said justices of the peace and commissioners of parochess, by the Committie of War within the Sheriffdome and presbytrie.

And lastly, it is appoyntit, for eschewing inconveniences ariseing by maisterless and idle beggares, that the act of parliament maid anent idle and maisterless beggares, discharging all and everie man to ressait, harbour, or to give almes to any except those of thair own paroche, be put to executione, conforme to the tenor thairof.

And ordaines the present act to be printed, and ane coppie thairof sent to everie paroche kirk, whar first it shall be read and thairefter battered on the kirk-wall, to be read by every persone, that none pretend ignorance thairof.

The Committie of the Stewartrie foirsaid, halden at Kirkcudbryt, upon the last day of September, 1640. Collonell, preses.

Act—Halyday and uthers.

The quhilk day, Johne Halyday of Fauldbey,[1] David Halyday of Marguillian, Johne M'Ghie in Barnebord, and James M'Connell of Creoches, becomes actit in the Committie's will for not subscribing of the generall band.

---

1 The family of Halliday hold the following tradition regarding the introduction of their name into Galloway.—A chieftain of the name of Halliday, who possessed an estate in the Highlands, had three children, two sons and a daughter; at his death the estate was divided equally amongst his three children, and shortly after that the daughter married a person of the name of Graham, who had been employed as a menial in her father's house. Her marriage greatly displeased her brothers, and a feud ensued in the family. Graham and his wife being joined by his friends and clansmen, proved too powerful for the two brothers, and they, fearing to remain any longer near their paternal residence, disposed of all their possessions and came into Galloway. One of the brothers settled at Glengap in Twynholm, and the other at Kulchronchie in Kirkmabreck; in both of which places, the Hallidays continued to reside until about about thirty years ago, when the descendants of the one brother left Glengap, and those of the other quitted Kulchronchie, both at the same term.

From the Register of Deeds it appears that in January 1587, John Mure of Cassincarrie wadset the half of his crofts and steadings, with house and pertinents of the Ferry of Cree, to John Halliday of Glen, for 200 merk scots. And in 1587, John Halliday having lent the sum of 100 marks to

Act contra M'Mollane.

Ordaines David M'Mollan in St. Johne's Clauchan, for his contempt to his captaine, minister and elderes, in not goeing furth to the armie, being enrolled, to pey presently fourtie punds, and to stay in ward in the tollbuithe of Kirkcudbryt, until the day of the rendevouez at Milnetoun of Urr, and then to march with the rest of the runawayes, and gif the said fyne of fourtie punds beis not peyit befoire he march, in that caice he shall pey ane hundred merks of fyne.

Act contra M'Guffok.

Ordaines John M'Guffock, runaway, to stay in ward, place foirsaid, until the foirsaid rendevouez, and then to march with the rest of the runawayes, and that his wyff get his haile guides, except sua much thairof as will carry him to the armie.

Commissar Depute.

Ressaivit by the Commissar depute, the rentalles of the pretendit bischopes' rentes and uthers unfreindes, lyane within the parochess efter specifit, viz :—within the paroche of Dalry, Balmaghie, Borge, Twenome, Kirkmabreck, Parton, and Girthetoun.

John Henderson, burgess of Kirkcudbright, got possession of twa buithes and ane chalmer, in that burgh, as security for repayment of the money.

The Hallidays of Glengap were firm adherents to the Covenant, and in 1685, David Halliday, portioner of Mayfield, Andrew M'Robert, James Clement, and Robert Lennox of Irelandton, having been surprised by Grierson of Lagg upon Kirkconnell Muir, in the parish of Tongland, were barbarously killed. In the same year David Halliday in Glengap, was shot by the Laird of Lagg, and the Earl of Annandale. Both these David Hallidays were interred in one grave in the church-yard of Balmaghie.

Act—Johne Somervaill.

The said day, anent the supplicatioun presentit by Johne Somervaill, expectant minister, shawing, that, whairas, at command and desyre of the presbytrie, he did serve the cure at the paroche kirk of Buittle, during the space of ane quarter of ane yeir, for the quhilk he hes ressaivit no satisfactioun, as the supplicatioun at lengthe beares. The said Committie, efter consideratioun had of the foirsaid supplicatioun ordaines that the said supplicator shall have peyit to him for the service foirsaid, by the parochinaires of Buittle, the soume of twa hundred merks monie, to be collectit to him by David Cannan of Knockes, off the crope and yeir 1640.

The Committie of the Stewartrie foirsaid, halden at the said burgh of Kirkcubryt, upon the first day of October, 1640.—Collonell preses.

Cannanes.

The said day Johne Cannan of Kirkennan, and Alexander Cannan in Braidlyes, becomes actit in the Committie's will, for thair not subscryveing of the generall band.

Act contra Halyday.

The said day, ordaines Johne Halyday of Fauldbey, for his contemptous lyeing out and not subscryveing the generall band, and for uttering of impertinent words, saying—That it was ane better world in the pretendit bischopes' tyme than now, when men are fyned; and, that, we will be glad to get thame agane yet: and also, for his misbeheavour to the table; to pey twa hundred merks of fyne to William Glendonying, and to lye in the stockes in prisoun untill the samyn be peyit, and to satisfie the kirk for his scandalous speiches.

The Committie of the Stewartrie foirsaid, halden at Drumfries, the fyfth day of October, 1640. The Collonell preses.

Act—Commissar.

The quhilk day, William Glendonyng, Commissar, desyrit the Committie to put pryces upon the boll of victuall pertaining to ante-covenanters, pretendit bischopes and uthers within the Stewartrie. The Committie ordaines him to uptak the pryce according to the feirs of the yeir.

Act—William Lindsay.

The quhilk day, the Committie ordaines William Lindsay to present the horss, allegit perteining to David Macbrair,[1] the next Committie day, and or-

1 About the year 1668, David M'Briar, an heritor in Irongray, who had been a member of Middleton's Parliament and accused his minister, Mr John Welsh, of preaching treason, became a violent persecutor. In a short time his affairs fell into disorder and he, being afraid that he would be imprisoned

daines the said David Macbrair to present the boy that delyverit the horss, or else to consygne for the horss j^{c} merks, and the horss to be delyverit to the said David.

James Gordoun.

The quhilk day, James Gordoun of Croftis, becomes actit to abyde the Committie's will for his lyeing out sua lang in not subscryveing the covenants.

The Committie of the Stewartrie foirsaid, halden at Drumfries, the sexth day of October, 1640. Collonell preses.

Thomas Thomsone.

The quhilk day, compeirit Thomas Thomsone in Richerne, and deponit, that he delyverit twa horss of David Macbrair's, to be keipit by him untill David

---

for debt, lived in concealment amongst his tenants. Whilst he thus lurked in Irongray, he was met by one John Gordon, a north-country merchant, who was engaged as an agent by a curate who had come from the north to Galloway. Gordon, observing M'Briar's melancholy and dejected appearance, immediately concluded that he was one of the covenanters who had been outlawed, and required him to go as a suspected person with him to Dumfries. This the other dreading imprisonment for his debts, refused to do; and Gordon having drawn his sword attempted to compel him. M'Briar either in endeavouring to make his escape or in resisting him, was run through the body and instantly expired. Gordon made no secret to the people in the neighbourhood of his having killed a whig; but, when they saw the body, they told him he had killed a man as loyal as himself, and having seized him, they carried him to Dumfries, where he was immediately condemned and executed.

came hame to the countrie, David repeying the horss' charges, and gif William cannot get baithe keipit to send the ane to Bargaltoun.

Act—William Lindsay.

Ordaines William Lindsay yet to keip the horss, untill the tyme that it be knowne whether it be convenient or not, and ordaines William to be peyit for the horss' charges, and that William keip the horss in good caice.

The Committie of the Stewartrie foirsaid, halden at Kirkcudbryt, the threttene day of October, 1640. Collonell preses.

Act—Volunteires.

The Committie ordaines that the persones who are appoyntit to contribute to the help of the volunteires and peyes not thair contribution at the rendevouez, shall pey thairefter the doubill; and ordaines the rendevouez for the volunteires, to be keipit at the Mylnetoun the xvij of this instant.

Answeris to the Artickles sent be the Committie of War of Kirkcudbryt.

1. As for the hows of the Threive;—ordaines the hows of the Thrieve to be flighted, and ordaines this to be done be Erlistone and William Griersone of Bargaltone, and recomends to the Committie to assist the samen.

2. Quhairas, we are informed that thair are sundrie persones amongst you enemeis and evill effected to the caus, and who neglectis and refuiss the publict orders given furthe for the good of the countrie, and does not pey nor contribut that quhilk is dew be thame in thair persones and estates for advanceing of the good of the publict. These does thairfoir give warrand and commissione to the Comimttie of War, to tak and apprehend all such persones and send thame heir, to the Committie of Estates, that they may be censured and punished answerable to thair deservings.

3. Item,—As for these that hes not subscryvit the generall band and hes peyit thair 10th and 20th pennies and uther impositiones, and who hes put furthe thair foote and horse, and hes assisted the common caus, and cumes in the Committie's will for censure;—remitts to the Committie of War to fyne thame, as the said Committie will be answerable, and to compt for thair fynes to the publict, and if they repyne, to put thame in prisone till they pey ther fynes or good securitie. And as for these that hes naither subscryvit nor will cum in will, but stands owt, they are to be fyned in ane heigher degrie answerable to thair caryege.

4. As for the wyffes and childrene of non-covenanters who hes contributed to the good of the caus, albeit thair husbands be averse thairfra;—ordaines the Commissar to intromit with thair Estates, resyrveand to the Committie of War to assygne to the wyffes and bairnes ane thirde pairt of the frie estaites

of thair husbands, whar the whole exceeds not ane thousand merks of rent, and if the rent exceed ane thousand merks, they are to meane thame to the Committie of Estaites, who shall give thame mantainance answerable to thair condition.

5. As for covenanters to whom thair is debtes awing be non-covenanters;—the creditores may not poynd at thair awn hand, but the Commissar must lift all; and the Committie of War shall tak tryall of all trew and lawful debtes, reallie owand to covenanters and freindes to the countrie and caus, and upon report of the Committie of War, under thair hands to the Committie of Estaites, and upon sight of the grounds of the debtes, with the rentall of the non-covenanter's lands, the Committie of Estaites shall give orders for thair satisfaction, efter tryall of the rentes and dew debtes of the non-covenanters.

6. Quhairas, Covenanters hes maid barganes and taken securitie frae non-covenanters for reliese of thair ingaegments;—it is ordainit, that, notwithstanding of the said rights or barganes, the Commissar shall intromit for the use of the publict; and these that hes maid such barganes or securities, must meane thamselffes to the Committie of Estaites. And efter tryall of the barganes and securities, the Committie of Estaites will give order to allow or dissallow of the said rights and barganes, as they find thame to deserve.

7. Item,—The Committie of War are requyerit to assist the Collectores and Commissares in everie thing necessar for the good of the publict, and for

the getting of all that is dew to the publict, and gif neid beis, to assist thame to poynd and distreinzie, as they shall be requyerit be the said Collectores and Commissares.

8. And lastlie,—Alexander Gordoun of Erlistoun, and William Griersone of Bargaltoun, are heirby requyerit to report ane exact accompt of thir haile artickles, betwixt and the last of this instant, as they will be answerable.

(sic subscribitur,) Argyll.
Cowpar.
Burghly.
Craighall.
J. C. Gaitgirth.
James Scott.
Thos. Patersone.

Letter anent the not out-cumeing of Volunteires.

Ryght Honourabill,—We have ressaived your letter the nynth of this instant, whereby you excuse the not cumeing out of gentillmen for assistance of thair countrie and cause now in hand, whereby it seimes you would have us to think thair is not anie weill affected within the Stewartrie and bounds of your Committie. But we are verie loathe to have such ane hard opinion of so manie gentillmen, but raither believes it to be from the slowness of these of the Committie thamselffes, who naither by thair example nor by thair dilligence in uther things does that quhilk we would expect. Thairfore we have resolved for such things as are exagitat concerning your beheaviour to requyer yourselffes of the Committie preciselie to be heir at Edinburgh, the day of this instant, sua that such ane cours may be taken, (since it cannot be done voluntarly,) as may maist conduce for the weill of the countrie and caus now in hand. And whosoever disobeyes this, our command, we will tak thame as unfreindes.

Ye are requyered to cume, prepared and resolved, to give ane exact accompt of your dilligence in the haile instructiones sent to you this whole tyme bygane, anent the affaires of the publict, and anent everie thing else quhilk were necessar and incumbent to you, as the Committie of War that in bounds, to have been done for the weill of the publict. So we expect ye will cume so prepared, to give ane accompt of everie thing quhilk can be demanded of you, as ye shall not neid to mak anie delay when you are cume heir, in answering and satisfeing of what ye are tyed to mak ane accompt of. So expecting your personall presence, with all possible diligence, immediatelie efter the ressait heirof, we rest your affectionate freindes,

Cowpar. J. C. Gaitgirth.
Burghly. James Scott.
Craighall. Edward Edgar.
Thos. Patersone.

Edinr., 9 Oct., 1640.

To the Ryght Honourabill, the Gentilmen, and uthers, of the Committie of War at Kirkcudbryt.

The Committie of the Stewartrie foirsaid, halden at the Mylnetoun, the xvij day of October, 1640. Collonell preses.

Logane.

Ordaines Irving of Logane, and David Cannan, to give in ane just rentall of all pretendit bischopes' and non-covenanters' rentes within thair boundes, under thair hands, at the Committie to be halden at the Threive, the xix of this instant.

Thomas Hutton.

Ordaines Thomas Hutton, and Gilbert M'Quhen of Netherthird, to doe the lyke within the parochen of Keltoun.

Act—Agnes Gordone.

The quhilk day, anent the supplicatoun given into the Committie, be Agnes Gordone, spouse to Robert M'Clellane of Nuntone, ante-covenanter, desyring thame to appoynt to hir ane competance out of hir said husband's estate. Taking to thair consideratioun the said supplicatioun; declares and ordaines the said Agnes to have the Kirkland of Dunrod, laboureing and manureing thairof, for ane pairt of ane competance of hir said husband's estate for hir and hir children.

Letter in favores of Gilbert Browne of Bagbe.

Ryght Honourabill.—For so much as we are informit that Gilbert Browne of Bagbe, who was in companie with the Erll of Nithisdaill in the hows, is apprehendit be the Lord Kirkcudbryt, and deteanit be him as ane prisoner, quhilk is alledgit to be contrair to the artickles of agrement betwixt us and the Erll of Nithisdaill Thairfore, these are to desyer you to tak tryall upon what grounds the said Gilbert is apprehendit and deteanit, and mak report thairof to us, to the effect we may tak such cours with him as shall be expedient. And so we rest, your affectionate freindes,

Cowpar. Murray.
Burghly. Thos. Patersone.
Seafort. Edward Edgar.
Robert Mour. James Scott.

Edinr., the 6 October, 1640.

To the Ryght Honourabill the Committie of War at Kirkcudbryt.

The Committie of the Stewartrie foirsaid, halden at the hows of the Thrieve, the xix day of October, 1640. Collonell preses.

Act anent the Threive and Lard Balmaghie.

The quhilk day, the Committie, in obedience of the warrand sent to thame from the Committie of Estaites, for flighting of the hows of the Threive; ordaines the said hows of the Threive to be flighted by the Lard of Balmaghie, as follows, viz.,—That the sklait roofe of the hows and batlement thairof be taken downe with the lofting thairof, dores and windowes of the samen, and to tak out the haill iron worke of the samen, and to stop the vault of the said hows. And with power to the said lard of Balmaghie, to use and dispose upon the tymber, stanes, iron worke, to the use of the publict; his necessar charges and expenses being deducted. And ordaines him, during the flighting thairof, to put sex musqueteires and ane sergand thairin, to be enterteanit upon the publict.

Act—Barscoib.

The quhilk day, the Committie of the Stewartrie of Kirkcudbryt, anent the supplicatioun presented by William M'Clellane of Barscoib, shewing that, where, he hes use for certaine friestane for building, and that he would buy frae the said Committie as manie as would serve him of the friestanes of the hows of the Thrieve, now ordainit to be flighted; as the said supplicatioun beares. The quhilk supplicatioun being heard, seen, and considerit; ordaines the said lard of Barscoib to tak as manie of the foirsaid friestane of the said hows, as will serve for his use, and to be in the Committies' will for the pryce thairof.

Letter to Ensigne Gibb.

I did heir at the Committie at Edinburgh, that they had written to the Committie of Galloway, answering to thair letter, that they had fund the Threive to be unprofeitable, giving present orderes that they should flight the samen. If they have desyerit you to cum out, that they might flight the samen, seing the warrand and taking the coppie thairof, signed under thrie or foure of thair hands. In doeing heirof, cum out with your garesone. Thir presents shall be to you sufficient warrand.

At Drumfries, the 17 October, 1640.

(sic subscribitur) Home.

Post.—Gif they give not this, stay still and let me heir from you shortlie, in all haist.

The Committie of the Stewartrie foirsaid, halden at the Mylnetoun, the xxij day of October, 1640. Collonell preses.

Letter anent the Out-putting of Volunteires.

Ryght Honourabill.—We find, to our great regrait, that, notwithstanding, we have so often and ernestlie recommendit the out-putting of the troupers for the last recerve, and these troupers of the first leavie that cam not furthe; and withall, invited all gentillmen and volunteires who would witness thair affection to thair countrie, now to cum furthe for the defence thairof; yet we have fund such a securitie and neglect bothe in putting furthe these that are dew to publict orderes, and in the volunteires, who are obleischit in honor and religione to join in this bussines, that we will be necessitat to tak harder courses and not lose anie more letters, sieing the samen avail nothing.—And sieing the estaite of the bussines now is such as the samen cannot admit any dispensatione; but, all that hes horss, without exceptione, must cum furthe upon thame, or send thame furthe with thair freindes or servands; and if any refusis ane of this twa, must not be countriemen, and will be repute as deserters of thair countrie and of this caus.

Quhairfore, these are to requyer you, that, with all possible dilligence, you cause put furth all these horss dew furthe of your schyer, be proportion, conforme to the former orderes; and what more horss are able for service, that the awners thairof cum furthe upon thame, or send furthe thair horss with uthers, fourneished with such armes as can be had. They shall have frie quarteres that are volunteires, and shall be frie of all toylsome dewtie, and the ordinar troupers shall have pey conforme to the orderes.

We are forced to tak more exact accompt how the orderes from this are obeyed, nor we have done in tyme past; and thairfore, we heirby requyer you, that thir orderes may be presentlie dischargit be you, befoir the 24th of this instant, or utherways, that you of the Committie of War give your personall appeirance heir, the said day, to render ane exact accompt of your dilligence and particular notes of these that are remise heirin. Quhairin if you failze be assured you shall incur the censure of this table. And these who shall be justlie dilated be thame, shall incur a censure to thair dishonor, besyde the taking of thair horss from thame. We have enacted this letter and will requyer and expect ane exact accompt thairof, the foirsaid day, or sooner if can be.

And howbeit, the Erle Argyll be now upon his march toward the army, yet, let that be no hinderance to the volunteires, for when they come heir thair shall be orderes given thame, and they shall have ane convoy till attend thame, till they cum to the army. Let not the present treatie stay anie from coming furthe, sieing the onlie meanes to obtean our wished desyers is the strength of our army at this tyme, and our peace and libertie does onlie, under God, depend upon the present strengthening thairof. And so we commit you to God. Your affectionate friendes.

(sic subscribitur)

Argyll.
Seafort.
Burghly.
Cowpar.
Craighall.
Caprintone.
J. C. Gaitgirth.
Edward Edgar.
James Scott.
Thos. Patersone.

Postscript.—You must have all your foote companies upon twenty-foure houres advertisment in readiness to march, for we know not how soon we be necessitat to call thame furth. Let exact and particular answers be maid heirof, and all the former instructiones sent to you anent the silver worke, lent monie, the peyment of your tenth and twentieth pennies, the provisioune for the sogers and all uthers recommendit to you. Efter the wryting heirof, we have resolved to caus the Erle of Argyll stay till all the voluntieres cum heir. If the late covenanters, or those who are double covenanters cum not furthe in proper persone, we must esteeme that they are onlie cum in for thair awn endes, without respect to the common caus.

Edinr., 12 October, 1640.

To the Ryght Honourabill, the Lord Kirkcudbryt, and remainent of the Committie of War within the Presbiterie thairof.

Letter anent the Delyverie of the Runawayes to Lieutenant Collonell Home.

Ryght Honourabill.—Whereas, Lieutenant Collonell Home hes gotten orderes from the Lord General, his excellencie, to march up with the south raigement to the army, with all convenient dilligence, and we are informit that thair is sundry runawayes from the army and his raigement within the bounds of your schyer, whereby the said raigement is greatlie wakinit, to the great prejudice of the publict. Thairfore, these are ernestlie to requyer and desyer you to mak searche and cause tak and apprehend all such fugitives, as weill frae the army as frae the said raigement, wherever they can be apprehendit within the bounds of your schyer, and delyver thame to the said Lieutenant Collonell, or one of his officers, for making up of the said raigement, and that within the space of foure dayes efter the ressait heirof, as ye will be answerabill, for this expedition cannot admit delay. By doeing whereof, ye shall give testimonie of your furtherance of the good cause now in hand, and the further oblige us to continue your affectionate friendes,

Argyll. Robert Mour.
Burghly. Dundas.
Cowpar. J. C. Gaitgirth.
Craighall. James Scott.

Edinr., 13 October, 1640.

To the Ryght Honourabill the Committie of War at Kirkcudbryt.

Letter from Lieutenant Collonell Home for the Runawayes.

Ryght Honourabill.—It pleissit the Lordes of the Committie of Estaites to send you this uther warrand for delyverie of all runawayes, bothe from our army and from our said south raigement, to Lieutenant Collonell Home, commanding the samen for the present.—Heirfoir, I humblic entreat your fortherance and care heirin; intending, on Wednesday next, to send ane officer to resaive the samen from you, hoiping ye will give thame a safe conduct to Drumfries, if neid beis. As for the garesone I laid unto the Threive, I have given the Ensigne orderes, on the sight and coppie of your warrand, quhilk ye did ressaive from the Committie of Estaites, the said double being subscryvit under your handes, that he shall render you the samen, to be disposet on as ye pleise. So, expecting your fortherance in all, as ye shall have me ever to remaine your humble servitor,

Home.

The Names of the Runawayes from Lieutenant Collonell Home his Companie out of Galloway.

Johne Bell, Johne Adair, Robert Phillop, Johne Makewn, Johne Andersone, James Makclanochen, Johne Makdougall, Patrick Cunnynghame, Edward Cawane, Adam Murrow, Charles Maxwell, Johne Smith, Hew Lyndsay.

Act—Johne Stewart.

The quhilk day, the Committie ordaines and admits Johne Stewart of Schambellie, as baillie within the baillerie of Newabbay, to doe, use and exercise all things necssar, and that appertains to the office of ane baillie, in sic caice. And ordaines the said Johne to contribute his best endeavours to the furtherance of the Commissar Depute in that bounds. And dispense with the said Johne his not goeing furth as ane volunteir at this tyme.

The Committie foirsaid, halden at Kirkcudbryt, the third day of November, 1640, be ane sufficient coram. Erlistoun preses.

Captaines.

The quhilk day the Committie ordaines that everie captaine, within this divisione, bring in all the runawayes to the next Committie day, being the xij of this instant.

Act—Johne Makmollan.

The quhilk day Johne Makmollan of Arnedarroch,[1] cautioner for David Makmollan, loyterar, presentit the said David and protests to be liberatit of his cautionerie. The Committie ordaines the said David to obey the former ordinance, given against him upon the last of September.

---

1 According to tradition the M'Millans in Galloway, are descended from some brothers of that name, who, having slain one of the Campbells of Argyle in a feud, fled from that county and settled at the Brockloch in Carsphairn, in which neighbourhood some of the family have ever since continued; indeed, at one time M'Millan was the predominant name in nearly all the upper district of the Stewartry.

Nesbit in his Heraldry says, "I have met with the old writs of Andrew M'Millan of Arndarroch, in the Barony of Earlston, amongst which I find his seal of arms appended to a right of Reversion in the year 1569. I find by their writs, they have been in Galloway in the reign of King Robert the III."

In the Register of Deeds there is an obligation dated 20 June, 1587, by which William Cunyingham of Polquharne, as principal, and Gilbert M'Caddam of Waterhead, as cautioner, binds themselves to pay to Johne M'Mollan in Brockloch, and Margaret Glendonyng his spouse, the sum of four hundred four score merks

Many of the M'Millans in Galloway joined the covenanting party and suffered for adhering to it. Mr William M'Millan of Caldow, in Balmaclellan, who had been licensed to preach by the presbyterian ministers in Ireland, was apprehended, conform to an order from the Privy Council, dated 20 July, 1676, and carried prisoner to Kirkcudbright, from whence he was removed and taken to Dumfries, where he was long detained prisoner. He was again apprehended in 1684, and suffered severly at that time The following extract, regarding him, is taken from the Burgh Records of Kirkcudbright:—

"At Kirkcudbright, the 13th day of November, 1676.

The quhilk day, Thomas Lidderdaill of St. Marie's Isle. Stewart deput of the Stewartrie of Kirkcudbright. presented to Samuel Carmont, ane of the Bailzies of the said burgh, ane order direct from the Lords of His Majestie's Privie Counsell, Quhairby the said Lords doe ordaine Maister William M'Millan, ane noter keiper of field conventicles, now prisoner in the tolbooth of the said Burgh of Kirkcudbright, to be transported to the tolbooth of Edinburgh. And for that effect, grants order and warrand to the Stewart of the Stewartrie of Kirkcudbright and his deputes, within the boundes of whose jurisdictione he is incarcerat, to tak the said Mr William M'Millan into his custodie, and to carrie him prisoner to the Sherff of the next adjacent shyer; and so furth from shyer to shyer till he be brought prisoner to the said tolbooth of Edinburgh. And ordaines the Magistrates

Act—Barquhillantie.

The quhilk day, the Committie ordaines Barquhillantie to intromit with the laird of Partone's crops, and to use and dispone thairupon, and to be answerable to the Commissar Depute.

Act—William Browne.

The quhilk day, the Committie ordaines the oxen allegit perteining to William Browne in Meikle Knox, to be apryset by Thomas M'Clellan of Collyn, and William Browne to give band to the Commissar Depute for the pryces, gif it be fund that the oxen perteines to him.

Act—Collyn.

The quhilk day, Thomas M'Clellan of Collyn hes peyit to the Committie twa hundred merks, in pairt

---

of Edinburgh to receive and detaine him prisoner therin until further order; as the said order subscribed by Mr Alexander Gibson, and datit at Edinburgh, the elevint day of October, now last bypast. Conforme and in obedience quhairunto, the said Thomas Lidderdaill, Stewart-deput, hes received from the said Samuel Carmont, Bailzie, the said Maister William M'Millan furth of the said tolbooth of Kirkcudbright; and the said Stewart deput hath delyvered him to William Herreis of Cloik; conforme to ane order direct from Robert, Lord Maxwell, principal Stewart And the said William Herries with his partie, is to convey the said Mr William M'Millan to the Sherff of Nithisdaill or his deput, who is the next adjacent Sherff; and to get ane ressait of him from them, for the said principal Stewart and his deput, their exoneratione. As witness their following subscriptiones.

Tho Lidderdaill
William Herries."

The Rev. John M'Millan, who was descended from a branch of the same family, was in 1701 appointed minister of Balmaghie. Being strongly attached to the more rigid principles of the presbyterian church and dissatisfied with the proceedings of the church courts, he protested against the defections and mismanagements of the church government, and afterwards became the first minister of the Reformed Presbyterians

of the peyment of the tenth penny of the parochen of Rerrick, for buying of clothe to the use of the publict.

Letter from the Committie of Estaites for Furneishing of Clothes and Schoes for the Sogers.

Ryght Honourabill,—We doubt not but you remember that, by the warrand and instructiones sent from this for buying of schoes and clothe for the use of the armie, ye were desyrit ernestlie to have all provisione that might be had of that kynd, heir against the viii of October, quhilk was the tyme intendit by us, and desyrit by our Committie at the camp, for the sending up to thame frae this of these commodities. We are presit daylie from the camp, to send thame the samen up with all possible haist, by reasone that all the poore sogers are almost perisched and in danger of thair lyves for want of schoes and clothes. Besyde the disgrace that our nation sufferis throw thair goeing naked in a strange countrie.

Quhairfore, we ernestlie entreat, that with all possible dilligence, ye will cause haist hither all the schoes and clothes, but especialie schoes, quhilk are or can be had within your bounds; for, if they come not with expedition the want of thame will lose all our sogers.

Next, we expected to have resaivit some exact acompt anent quhat silver worke is to be had in your bounds, and that the samen befoire this tyme should have been sent heir and delyverit for the use of the publict; upon quhilk we entreat may yet be done with dilligence. And also, that, the voluntar contribution that is, or may be had or collected within your bounds, for advancement of the good cause, may be haistened with all dilligence.

You are lykewayes to requyer the Collector Deputes and Commissar Deputes within your bounds, to come heir, with all expedition, and render ane accompt of the dilligence uset by thame of the uplifting of the tenth penny, twentieth penny, contribution, and the rentes of bischopes, ante-covenanters, papists, and uther unfriendes, within the bounds allowit to thame. Requyer thame also, to bring with thame thair compts, together with all the money they have medlet with for the use of the publict. To the effect that thair bygane intromissiones and dilligence may be cleirit, and that they may resaive orderes for thair cariage and dilligence heirefter. Let this be ane advertisement to thame; and gif they be not requyerit by you to come so prepared, the blame shall be imputed to you; and gif they faill, being requyerit by you, we shall tak cours for thair censure.

We expect and desyer ane particular and exact answer of the haile premises, preciselie, betwixt and the first day of November; at quhilk tyme we will call all your committies, and uthers to whom this is directed, by your names, and gif they naither appeir nor send we will tak hardlie of it. But, in the meane tyme, we wish all the schoes and clothe may be sent with all expedition.—Your affectionate freindes

P.S.—We entreat that ane band of the tenth penny be sent heir with all expedition.

(sic subscribitur,) Argyll.
Mar.
Burghly.
Cowpar.
Murray.
J. C. Gaitgirth.
Richard Maxwell.
Thos Patersone.
James Scott.

Edinr., 23 October, 1640.

To the Ryght Hon. the Lord Kirkcudbryt and Committie thair.

Letter from the Estaites in favores of Johne Maxwell of Newlaw.

Ryght Honourabill,—Whereas, Johne Maxwell of Newlaw hes represented to us his kyndlie interest to the Abbacie of Dundrennan, bothe as heritor of a great pairt of the lands and teinds of a pairt of the patrimonie thairof, and as uplifter and feuar of the rest of the samen. And as we resolve not willingly to wrong the publict, so we doe not love to wrong onie man in thair ryght or kyndlie possessione, so being the publict be not prejudycit In respect whereof, we have condiscendit to give him ane factorie, upon sufficient cation fund by him for being comptable to the publict.

And, in respect we resolve to doe nothing in that kynd bot with deliberatione, thairfore, and leist any sould be interest heirin, we resolve to keip the commissione undelyverit till we heir from you, gif the granting of this warrand to the said Johne Maxwell may be prejudycial to any uther pairtie's ryght.—Quhairin, gif thair be no uther concerned nor prejudycit heirby, we resolve to give him his warrand, sieing the publict is securit, by his band and caution, for compt and reckoning.

We expect your answer hereon, betwixt and the tenth of November next; and in the meane tyme, because William Glendonyng hes a factorie for all bischopes' rentes within your Stewartrie, you shall desyer him to forbear the rentes of the said Abbacie till we give farther orderes; whereanent thir presents shall be ane warrand. And so we commit you to God, Your affectionate freindes.

(sic subscribitur,)

Burghly.
Cowpar.
Dundas.
Rt. Mour.
Edward Edgar.
James Scott.
Richard Maxwell.

Edinr., the 19 October, 1640.

To the Ryght Hon. the Lord Kirkcudbryt and Committie of War of the Stewartrie thairof, and to Wm. Glendonyng, Commissar thair.

The Committie of the Stewartrie foirsaid, halden at Kirkcudbryt, the fourt day of November, 1640. Collonell preses.

Act—Mr David Leitch.

The quhilk day, anent the supplicatioun given in at the instance of Mr David Leitch, Minister at Rerrick, desyreing ane ordinance for peyment to him of his ordinar steipand, conforme to use and wont. The Committie ordaines the heritores and uthers, quho are adebtit and hes bene in use of peyment to him of thair teindes, within the said parochen of Rerrick, to redelie answer, content, obey, and mak tkankfull peyment to the said Mr David of his steipand, this instant crope of 1640, as they have bene in use of peyment befoire.

Act contra Captaines of Parochess.

The quhilk day, the Committie ordaines the Captaines of the parochess, at the next Committie day, to give in ane perfect inventar of non-covenanters' cornes, goods and geir within thair parochess; and that they caus apryse the samen be honest men; and to desyer the said non-covenanters, to whom the samen apperteines, to cum and find sufficient suretie that the samen shall be furth-cumand to the publict.

Act—Mocherome.

The qubilk day, ordaines William Glendonyng of Mocherome to citate William Gordone in Nether-corsok, to the next Committie day, to answer at the instance of Johne Sinklar in Knockgray, personallie or at his dwelling place.

Act—Lennox of Callie.

The quhilk day, ordianes Johne Lennox of Callie to citate Johne Muligane, to the next Committie day, at the instance of Andro M'Chesnie; and that they find suretie to uthers, as accords of the law.

Act—Captaines.

The quhilk day, ordaines the Captaines of the parochess to send their constables, with twa sufficient witnesses, to rype throw the parochess for suspectit gudes.

The Committie of the Stewartrie foirsaid, halden be ane sufficient coram, at Kirk-cudbryt, the twellff day of November, 1640. Collonell preses.

Letter from the Estaites, anent Watching for Runawayes, and anent thair Ressettares.

Ryght Honourabill,— Notwithstanding of the strict actes maid anent runawayes and fugitives from thair cullores, and the frequent and ernest desyring to put the samen to exact executione, and for keiping of passages in all pairtes of the countrie, for catching and apprehending thame; yet they daylie disband in such multitudes, that, within a schort space our army be all apeirance shall melt away, to the schame and ruine of this countrie and caus now in hand. And thair is nothing, that hes so much occassioned and fosterit the disbanding as the not putting of the actes, sent from this to dew executione, anent the searching and apprehending of runawayes, maisterless and idle men, and keiping of all passages straitlie, conforme to the former actes.

These are thairfore to requyer you, as ye will be answerable to the Estaites of this kingdome upon the hiest perall, as ye love the countrie and success of the bussiness now in hand, that, immediatelie on the receipt heirof, you put strait watches to all pairtes within your boundes, quhair onie uses or may cum from the south pairtes; and that nane be sufferit to cum throw your bounds without a lawfull pass, subscryvit be the Lord Generall or Committie of Estaites. And if anie shall be fund cuming from the army without a lawfull testimoniall, as said is, of whatsoever qualitie they be, without exceptione, shawn and approven be these intrusted to the keiping of the watche; ye are heirby requyered, aither to keip thame in firmance till ye acquaint us thairof, or utherwayes to send thame bak to the army, as ye will be answerable.

Let these presents certifie you, that if this order be not exactlie and punctuallie observit be you, ye shall be called and censured for your negligence answerablie, as so great a neglect shall be fund justlie to deserve. And be assured that if anie escape your watche, wanting a testimoniall, and shall be fund in anie pairt of the countrie, the way that they have past thither shall be tryed and baktryed, and the blame and censure of thair escape shall lye upon you.

We recommend lykewayes to you, the strict executione of the said former actes against runawayes, idle and maisterless men, till farther orderes; for God willing, ye shall have schortlie printed actes and orderes for remeid of this great and fearfull evill.

Remitting the premises to your care and dilligence, as ye will be answerable as said is, we continue your affectionate freindes,

(sic subscribitur,)

Burghly.
Cowpar.
Capringtone.
Gaitgirth.
Mr Wm. Moir.
Thos. Patersone.
Richard Maxwell.

Edinr., 2 Nov. 1640.

For the Ryght Honourabill, the Lord Kirkcudbryt and Committie of War of the Stewartrie thairof.

Act—David Macmollan.

The quhilk day, David Macmollan, loyterar, being convenit for saying, that Galloway should not keip Mr Hew Hendersone, his minister, and him bothe; depones, out of his awn mouth, that if Mr Hew did not freithe him of ane sclander laid upon him be the parochen, in reporting him to be the persone that said he wald be drunken with Armacannie, that the said minister and he should not keip Galloway. The Committie ordaines, for the said caus and uthers foirsaid, quhairin he was decernit in a fyne befoire, to pey for all ane hundred merks, and to stay in ward untill the samen be peyit, and to sit the morn in the stockes betwixt ix and xij houres, with ane paper on his heid beirand this device, AS ANE LOYTERAR, with the foirsaid speach of his minister.

Knockschene.

The quhilk day, Johne Gordone of Knockschene undertakes to produce his sone, David Gordone, runaway, to Erlistone, at the next Committie day.

David Cannan producet Johne Campbell, runaway.

James Makconchie producet Johne Makconchie, runaway.

Johne Boddan producet Johne Crosbie, runaway.

Robert Makillvenie producet Johne Makilvenie, runaway.

Glaisteres producet Johne Birkmyer, runaway.

The Committie ordaines Robert Prymrois to produce James Maknaught.

Act—Robert Glendonyng.

The quhilk day, anent the desyer proponit be William Glendonyng, Commissar Depute, desyreing that Robert Glendonyng, notar, his eldest brother, might be approven be the Committie as his depute, in manner and to the effect conteinit in his commissione. The said Committie approves of the said desyer, and admits the said Robert Glendonyng, and ordaines him to be thankfullie obeyit as weill as the principall, and that under the paine of censure to be incurred be the disobeyares.

Act—Captaines.

The quhilk day, the Committie ordaines the Captaines, *sicut ante*, to produce ane inventar of the non-covenanters' rentes, gudes and geir, and the non-covenanters to find cautione, conforme to the former act.

Act—Barquhillantie.

The quhilk day, the Committie ordaines Barquhillantie, *sicut ante*, to intromit with the lard of Partone's crope, use and dispone thairupon, as said is, to the use of the publict.

Apprysares of Bagbie's Cornes.

The quhilk day, compeirit Johne Martene in Newtone, Gilbert Raen in Bischoptone, William Raen, thair, and Johne Robesone, wha were appoynted to have apryset Gilbert Browne of Bagbie his crope in the Nuntone; depones upon thair great oath, according to thair knawlege, thair are in the barnes and barne-zeard of Nuntone, perteining the said Gilbert Browne, thrie score bolls aites and ten bolls beir.

Act—Knockschene.

The quhilk day, the Committie ordaines Johne Gordone of Knockschene to produce ane act, allegit purchasit in his favores be Mr John Diksone and uthers frae the Committee of Estaites, anent the sub-seryveing of the general band, and that within fourtene days efter the date heirof, utherwayes his fyne shall be exacted.

The Committie foirsaid, halden be ane sufficient coram, at Kirkcudbryt, threttene day of November, 1640. Collonell preses.

Act for Peyment of the Tenth and Twentieth Penny.

The quhilk day, the Committie ordaines, that upon the first day of December next, the Captaines of the of the parochess inbring to the generall collector, the haill tenth and twentieth penny rentes, yet restane within the boundes unpeyit, and that under the paine of censure; and where they cannot get peyment frae the persones, debtors, to poynd thame thairfore, viz.—the ox for ten punds, and the kow for ten merks, and to be of the best gudes they can get; and to bring the said gudes, sua to be poyndit, to the Committie the said day, where they shall be taken for sufficient peyment.

Act—Sinklar contra Gordone.

The quhilk day, anent the supplicatioun presented be Johne Sinklar in Knockgrey,[1] schawing that one Johne Scott, runaway, cam to his hows and stole furthe of his kist, thrie double peices and sextene punds of whyte silver; and that thairefter, William Gordone in Nether-corsock reset the said runaway, and heiring that he had stollen monie he tuik the samen frae the said runway and put him away; desyreing that he might be repaid by the said William what he lost be the said runaway. The Committie, having causit citate the said William Gordone befoire thame, for the same caus, and he not compeirand, being called; and it being notorlie knawn that the said William reset the said runaway and tuik a pairt of the said monie frae him, as he declarit to his captaine. The

1 A family of the name of Sinclair were proprietors of the lands of Auchenfranko at a very early date. In Pitcairn's Criminal Trials it is stated that on "January 11th, 1542-3, John Maknacht of Kilquhannite, (being then at the horn,) found surety to underly the law, at the next Justice aire of Kirkcudbright, for art and part of the cruel slaughter of William Sinclair of Auchenfranko. (April 17, 1543,) Andrew Herys, brother of William, Lord Herys, became surety for his appearance to answer for the said crime" According to tradition, Sinclair was murdered at the Netherplace of Urr, and the field in which the murder occurred still retains the name of the Sinclair Yard

In the Register of Deeds there is a copy of agreement, dated 1st July 1585, by which Robert Herries of Mabie, to whom the lands of Auchenfranko had been sold by William Sinclair elder, for two thousand marks, grants letters of redemption to William Sinclair younger. This agreement was cancelled and another made in 1587, in which the sum given by Herries is stated to be four thousand marks scots.

said Johne Sinklar deponit, be his athe, that the said Johne Scott, runaway, stole from him the said thrie peices and xvi libs monie. And having taken the haill premises to consideratione, ordaines the said William Gordone to content and pey to the said Johne Sinklar, the said soume. And ordaines George Glendonyng, Captaine of the parochen of Partone, to sie the said Johne Sinklar compleitlie peyit of the said soume, or else delyver him as meikle of William's gudes as will pey him.

Act contra William Gordone.

The quhilk day, the Committie ordaines the said William Gordone to content and pey to William Glendonyng, Commissar Depute, for resetting of Johne Scott, runaway, the soume of j^c^. libs.

Act contra Johne Makartnay.

The quhilk day, the Committie, for the ungodlie words and filthie speiches utterit be Johne Makartnay[1] against the umq^le^. Minister of Urr his wyfe,

---

[1] "The Macartneys are said to be descended from Donough Macarthy, younger son of the ancient and warlike Irish family of Macarthy Moore. In the beginning of the 14th century, their son Donough (or Daniel) having served Edward de Bruce in Ireland, went, after the Battle of Dundalk, to king Robert de Bruce, in Scotland, whom he also served in his wars, and from whom he obtained a grant of lands in Argyleshire. His descendants, being dispossesed of their lands, removed into Galloway, and, with their bow and sword, acquired the lands of Loch Urr, Macartney, and others. In their castle of Loch Urr, Sir Christopher Seaton, brother in law to Bruce, was taken by the English, in 1306, (being betrayed by one M'Nab,) and carried to Dumfries, where he was executed This family soon spread into several branches, in the barony of Crossmichael—the greater part of which they held in feu from the college of Lincluden till the Reformation, when the Viscount Kenmure obtained, from the Crown, a grant of Superiority

efter tryall taken be the said Committie thairof, ordaines the said Johne, upon Sounday next, to ryse out of his seat efter sermone and confess his fault in declameing, by words, of the gude name and fame of the said relict and to crave first God's mercy for offending him, next the said relict's for the offence done to hir, and then the parochinares' for his evill example, and then his minister to resaive him and the said relict to tak him be the hand, and to pey to the Commissar Depute for his fyne fiftie merks monie.

---

and property of the said college. The family divided into three principal branches, viz, Meickle Leaths (Buittle,) Auchenleck (Rerwick,) and Blacket (Urr.) From that of Blacket was descended General Macartney, (1729,) and also the celebrated Earl Macartney; and from Auchenleck is descended Alexander Macartney, Esq., of Ballocco, (Rerwick).—HISTORY OF GALLOWAY."

In the Register of Deeds of the Stewartry of Kirkcudbright, there is an obligation, dated 26th June, 1585 by which Peter Cairns of Kip as principal, and Johnne Lewres in Auchencairn as cautioner, bind themselves to content and pay to Johnne M'Cartney of Lethis, alias callit Wattis' Johnne, each year for four consecutive years, four bolls meill and twa bolls of beir, greit messor, together with fifty merks monie, and that as compleit peyment of twa hundred and fifty merks,—the tocher which Peter Cairns was to have given with his daughter Issobell Cairns. The same John M'Cartney is witness to a deed, dated 24th October, 1587, in which Johne Stewart, feuar of the twenty shilling lands of the fourty shilling lands of Auchenleck, grants him to have sold the said lands to Bratill M'Cartney, son and apparent heir to Patrick M'Cartney in Auchenleck, for the sum of £207 13s 4d. scots. And on the same date John Stewart sells the five shilling lands of Auchenleck to John M'Cartney, son of Adam M'Cartney, who then occupied the lands, for the sum of fiftie four punds and fourtie pence.

The M'Cartneys acquired the lands of Blacket by intermarriage with a family of the name of Hillow, who were proprietors of Hillowton. In the Register of Deeds Robert M'Cartney, brouster in Hillowton, is witness to an agreement, dated 2nd January, 1587, in which Eupheam Gordoun, daughter and one of the heirs portioners of John Gordoun of Blacket, and Archibald Hillow of Hillowton her spouse, having borrowed the sum of six hundred and fifty marks from William Grahame, grants him security on the eleven shilling and six penny lands of the ten mark lands of Blacket and the eleventh part of the mill of the same.

George Macartney of Blacket having adhered to the Covenanting party suffered much during the persecution; he was imprisoned for above six years, his estate seized and his lands laid waste.

Instructiones frae the Committie of Estaites,

To the Commissares and Collectores throw the haill schyers of the Kingdome, which the said Collectores are hereby obleist to execute and discharge in all poynts as they will be answerable to the Committie of Estaites.

1. The Collectores, ilk ane rexane within thair own divisione, are obleist to bring in the haill tenth penny and uther dewties within the bounds quhilk cumes under the compass of thair divisione.

2. Everie Commissar, within his own bounds, is obleist to intromet with, uplift and be comptable to the Committie of Estaites and Generall Commissar for the rentes of all bischopriks, ante-covenanters, papistes, and uthers unfreindes within thair divisione. And siklyke, of the particular soumes, rentes and estaites propeilie perteining to a bischope's proportione. And siklyke, they are to be comptable for the moveable gudes, cornes and uthers, perteining to the said papistes, bischopes, ante-covenanters, except such as hes particular warrand for intrometing with the particular estaites and rentes of ane particular ante-covenanter, papist or bischope, and quhilk particular warrands the Commissares are not obleist to acknowlege, unless the samen be schawn and the coppie thairof delyverit to thame for thair exoneratione. They are also, heirby warrandit to uplift the quietus of all testaments frae the Commissares and thair phiscalls, and all uther ordinar dews, and to call thame to ane accompt for the samen.

3. They are obleist to intromet with and uplift the said rentes for this present cropt, and for all bygane croptes, yeires and termes, since the lawfull discharges grantit be persones having ryght thairto; and they are obleist to call for a sight of the said discharges and tak coppies thairof, quhilk coppies they shall give in with thair accompts, and must compt for all termes subsequent to the said discharges. Lykeas, they are heirby warrandit to call the intrometers with anie of the said rentes and uthers foirsaid, and mak thame comptable of thair intromissiones.

4. The Commissares are obleist, with all convenient dilligence, to give and delyver to the Committie of Estaites ane perfyte and particular rentall of the said teinds, benefices, patrimonies, and uthers whatsumever, perteining and dew to all ante covenanters, bischopes, or bischopriks, papistes, and uthers unfreindes within thair bounds; together with ane inventar of all thair moveable gudes, debtes, soumes of monie, fermes, and uthers perteining thame, sua far as they can get knawledge of; together with ane rentall of all rentes, customes and uther dewties peyable to the King's Majesty, within thair divisione. Which rentalles of ante-covenanters, papistes and bischopes' landes and gudes they are obleist, as weill with suche as they are to intromet with thameselffes, as for suche rentes as thair are particular warrands given in for; to the effect the Estaites may knaw what is dew to the publict throw the whole countrie.

5. The Commissares are to sell and dispone upon all the said rentes and gudes quhairwith they shall intromet for the use of the publict, and to mak monie thairof with all expeditione, which they must bring to the Commissar and generall Commissar or his deputes, for the use of the publict.

6. The said Commissares are obleist to intromet with and uplift the rentes of all uther benefices formerlie perteining to anie pretendit deaneries, or anie uther benefice whatsumever, suppresit and dischargit be the late actes of Parliament or Assemblie rexane; and to give up rentalles thairof and be answerable thairfore, in the same manner as is above mentionit.

7. The Commissares and Collectores rexane, are obleist to tak the advyse and requyer the concurs and assistance of the Committie of War, where thair severall charges and employments lyes; as also, to requyer the assistance of all Collonells, noblemen, gentillmen, burrowes and uthers rexane within the said boundes. Quhilk Committies of War, collonells, noblemen, gentillmen and uther persones, are obleist to give thair reall assistance to thame in everie thing necessarie, concerning the charge and employment foirsaid, as they or onie of thame shall be requyerit for that effect, and for the better furtherance of the service of the publict. The foirsaid persones are obleist to ryse, and caus the countrie ryse and assist, and send out pairties in armes or utherwayes, as occasione offers, for doeing of executione contra onie persone quhatsumever, who shall be refranares or

remise in peyment of what is dew to be uplifted be the said Collectores and Commissares, conforme to the commissiones and to thir present instructiones. And quhilk persones, who shall so ryse and be sent furthe for assisting of the said Commissares or Collectores, shall have libertie of frie quarteres upon the said persones, refuisers or delayers to mak peyment or to give obedience in manner foirsaid.

8. In caice anie Noblemen, Committie of War, Collonell, or uther persone whatsumever, shall refuis thair assistance, or to ryse to doe executione, in maner foirsaid, the said Collectores or Commissares shall have power heirby, to caus summond and charge the said refuisers and disobeyares, in assisting or ryseing as said is, to compeir befoire the said Committie of Estaites, to be censured for thair neglect and contempt, and that to anie day or diet the said Commissares or Collectores shall pleis to charge thame to.

9. The said Commisioners and Collectores shall be obleist continuallie and frequentlie, as they are able to get in monie or to get cornes or gudes converted into monie, to send or bring the same to the Commissar or Collector General, or their deputes, residing at Edinburgh, to be employet for the use of the publict; and if they keip up monie efter thair resait thairof, they shall be censured as abuisers of the trust and charge committed to thame.

10. The said Commissares and Collectors, at thair resaiveing of thair warrand and instructiones, shall be obleist to give thair oathes, *de fideli administratione*

in thair said offices, in presence of the said Committie of Estaites. As also, at the ingiving of thair accompts, they shall be obleist to subscryve the samen accompts, according to thair intromissione, and that they have omittit nor concealit nothing to the prejudice of the publict.

11. The Committie of War in ilk divisione, are heirby recomendit to tak notice, oversie and advert to the caryege and dilligence of the Commissares and Collectores in thair severall employments, and if they be negligent in thair dilligence, to reprehend thame or delate thair faults or miscaryege to the Committie of Estaites.

12. It is appointit, that the Collector of the tenth penny within everie presbit: rie, shall be Collector also of the twentieth penny, and bothe to be collectit together. Lykeas, these gives thame full power as weill for collecting the twentieth penny as the tenth penny.

13. It is also appoyntit, that the Committie of War in everie presbiterie or schyer, where thair valuationes are not as yet given in, call the valuers to ane accompt of thair dilligence and sie the valuationes exactlie set downe, conforme to the Act of Parliament, viz.—be the heritores' subscriptiones upon thair conscience and credit, or be valuers appoyntit for that effect, upon thair conscience and credit, that the samen is just and trew, according to thair knawledge and the best informatione they could get.

14. It is appoyntit, that all rentallers be valued as weill as the heritores, and that the full worthe of the land be valued as gif the samen were not rentalled, and that the valuationes be given up be the heritor or rentaller upon thair oathes, as said is. The heritor to pey the tenth and twentieth penny for his rentall dewtie, and the rentaller to pey the tenth and twentieth penny for the rest of the worthe of the land.

15. The said Commissar deputes are obleist and hes heirby warrand to convein the schyreffes, baillies, and thair deputes, and uther intrometers with His Majesty's rentes and the Prince's rentes, for the production of the last acquittances, and to mak compt and peyment sensyne, conforme thairto, that the samen may be brought into the Estaites and Generall Commissar, with all expeditione.

The Committie foirsaid, halden be ane sufficient coram, at Cullenoch, the xxiiij day of November, 1640. Erlistone preses.

Act for Inbringing of King's Rentes.

The quhilk day, the Committie ordaines, that ilk Captaine within thair severall parochess, caus uplift the Lardner mart ky, conforme to use and wont, and also to caus all that are adebtit to the King or Prince any rentes to bring in the samen with the last discharge, under the payne of fourfaulding thairof.

Act anent Baggage Horss.

The quhilk day, it is ordainit that the parochess that hes not brought in the baggage horss as yet, inbring the samen the next Committie day, or else to schaw ane resonable caus why the samen should not be done, with certificatione if they failzie, the Committie will give power to the Captaines of the parochess to poynd for the double.

The Committie of the Stewartrie foirsaid, halden be ane sufficient coram, at Kirkcudbryt, the first day of December, 1640.—Erlistone preses.

Letter for putting in executione the Actes against Runawayes and thair Ressettares.

Ryght Honourabill,—You are to resave heirwith the number of seventene printed actes against runawayes. Lykewayes, resave the number of four printed instructiones, quhairby ye will persave what is requyred of you, for the good of the publict.

We desyer that bothe the actes and instructiones may be publischit and put to exact executione throw your haill bounds, with all dilligence. We neid not be more particular, for these actes and instructiones will inform you sufficiently what is to be done be you, only we do ernestlie entreat and do certainlie expect the reall effares of your dilligence in all thir particulars, as tending to the use of the publict. And at the expyering of everie ane of the dyets prescryvit be thir instructiones, we will expect ane exact accompt of your travells and dilligence; for truelie we will be necessitat and resolve to cause execute the last certificate of the instructiones upon such as shall be fund remis. General answers to the poynts of thir instructiones will not suffice nor satisfie; thairfore everie poynt must be particularlie and punctuallie answerit and performed, for the necessitie and danger of the present tymes requyer a reall performance be everie man; and such as are not reall must be distinguished and taken notice of.

So, resting upon the assurrance of your faithfull care and dilligence heiranent, we commit you to God.
Your affectionate freindes,

(sic subscribitur,) Argyll.
Amond.
Montrose.
Eglingtone.
Cowpar.
Burghly.
Craighall.
S. P. Murray.

Edinr., 18 Nov., 1640.

For the Committie of War at Kirkcudbright.

Instructiones

Sent be the Committie of the Estaites of Parliament to the whole Schyres, Committies of War, and Burghs within this kingdome, the 16*th November*, 1640.

1. First, resave heirwith the actes against fugitives and runawayes and thair resettares, which must be proclaimed at everie mercat croce the first mercat day, and in everie kirk the first Sonday efter the receipt thairof; and for this effect thair is as monie actes sent to you as thair are paroche kirkes within your boundes, bothe to burgh and land, which actes you must send to everie kirk.

2 Secondlie, thir actes, as also the former actes against fugitives, maisterless men, and those who travell without pass must be put to dew executione, conforme to the tenores thairof; and all fugitives must be apprehendit and punished conforme to the actes, and the rest sent to Edinburgh within fyftene dayes efter the receipt heirof. Lykeas, strict cours must be taken in everie place for keiping of all heighwayes and passages for apprehending of all runawayes.

3. Thirdlie, all the clothes and schoes in each presbiterie and burgh alreadie providit for the sogers in the armie must be sent to Edinburgh, or to the camp, within four dayes efter your receipt heirof; and orderes must be given for making all the schoes and buying all the clothe that can be had in your boundes, which must be prepared and sent to the

armie with all possible dilligence; and at the delyverie thairof, you must give order to get the Commissar's ticket of the receipt of the samen, for keiping of a ryght compt, utherwayes what you send and delyver will not be allowit be the publict.

4 The Committies of War and Magistrates of Burrowes must send to the Committie of Estaites at Edinburgh, ane exact roll of the names of all ante-covenanters, papistes, and uther enemeis to the common caus within thair bounds; together with ane rentall of all thair landes, tythes and rentes; and ane inventar of all thair bands, soumes of money, moveable gudes, cornes and uthers perteining to thame, or to any bischoprik or bischopes within thair said bounds; together, also, with ane roll of the names of suche as profess to be covenanters and yet does not reall dewtie, and of the names of all uthers who are suspected not to be reall freindes to the common caus. And all this within xx dayes efter thair receipt heirof.

5. The said Committie of War, as also all Collonells, Noblemen, Gentillmen, Magistrates of Burrowes and uthers, must assist the Commissares and Collectores in everie thing conforme to the said Commissares and Collectores thair instructiones, and power given to thame in thair severall offices.

6. All the Commissares and Collectores must presentlie cum to Edinburgh with thair accompts and resaive new orderes and instructiones; and the Committees of War must requyer thame to come for

that effect. And if thair be anie pairt of the countrie where thair is not Commissares and Collectores established, the Committie of War must nominate thame and send thame to Edinburgh to get thair warrands, and this within aught dayes efter the receipt heirof.

7. That all the valuationes be closit, perfyted and and sent to Edinburgh, where the samen is not done alredie, and that within xv dayes efter the receipt heirof.

8. That all the tenth pennies and twentieth pennies be presentlie collected and sent to Edinburgh, except what is alredie peyit be publict order from the Committie of Estaites or Collector General; and the Committies of War are heirby requyered to assist the samen, and this within xx dayes efter the receipt heirof.

9. That the Committies of War and Magistrates of Burrowes recommend to all the ministeres within thair bounds, to be ernest in exhorting thair people to give in thair voluntar contributiones, which must be sent to Edinburgh with all dilligence, for advanceing of the gude caus. And that report be maid of thair dilligence, under the hand of each minister, within a month efter the receipt heirof.

10. That the Committies of War and Magistrates of Burrowes, rexane, do dilligence for sending the whole silver worke within thair bounds to Edinburgh, conforme to the printed instructiones thairanent; and that they charge befoire thame everie particular persone who are thocht to have silver worke, to de-

lyver the samen upon securitie to the use of the publict; and suche as compeirs not, or refuise to delyver what they have, to charge thame befoire the Committie of Estaites at Edinburgh; whereanent thir presents shall be ane warrand. And all this must be compleitlie done within a month efter the receipt heirof.

11. That the whole people in the kingdome, as weill to burghe as land, be drylled and exercised frequentlie; and this is requyered to be done be the Collonells, muster-maisters in each schyer, who hes former commissiones for that effect, which the Committie of War are heirby requyered to assist, and who are heirby requyered to send the muster rolles of thair whole men and armes, as weill horss as foote, within thair bounds, to the Committie of Estaites, and that within a month efter the receipt heirof.

12. That the said Collonells and Commanderes of each schyer, and the Committies of War tak present tryall, within thair bounds, of those of the first leavie, as weill of the fourth man as the eighth man, and of the trowpers, at twa thousand merks of rent, that were not put furthe to the armie, according to thair proportiones; and to tak a list of what is restane not put furthe, aither horss or foote; and to cause presentlie furneish thame with armes and uther necessares; and to tak assurance that they may be reddie upon two dayes advertisement to cum furth, with fourtie dayes lone, and this without prejudice of thair fynes for not cumeing furthe in due tyme.—

Lykeas, the said Collonells and Committies of War are heirby requyered to send a list and roll of the said horss and foote, yet restane, not cum furthe, to the Committie of Estaites, with thair names be whom they are due, and that within a month efter the receipt heirof.

13. As for the last recerve of the tenth man and ane trowper horss for everie sex thousand merks of rent, the Committie of War, Collonells and Commanderes, are heirby requyered to put thame at once upon foote, and to sie thame sufficientlie armed, and to tak assurance that they may be reddie to cum furthe upon advertisement, and to send to the Committie of Estaites a roll of the number, bothe of horss and foote, which may be outreiked, according to the proportion foirsaid, for the foirsaid recerve, furthe of eache schyer and division, and this within a month efter the receipt heirof.

14. All the volunteires who were reddie and did offer thameselffes to cum furthe in October last, and all uther gentillmen who have anie able horss and who affects this caus, are heirby ernestlie requeisted to be in reddiness upon the next advertisement.—And it is declared that anie volunteir who pleiseth to cum or send out, shall have answerable deduction aff thair proportion of horss for the recerve, according to a trowper for each sex thousand merks rent, provyding that, befoire they desert thair service, they be obliged to furneish thair due proportion of horss, according to thair stent.

15. Item,—that a perfyte roll be sent to the Committie of Estaites at Edinburgh of the names of the whole persones that are resaved and sworne upon each Committie of War, and the name of your clerk in each division, and this within eight dayes efter the receipt heirof.

16. It is heirby declared, that when anie of these who are of the ordinar number of the Committie of Estaites shall happen to be abroad, in anie part of the countrie, that they shall have place and voyce as ane of the ordinar number of the Committie of War, in the division where they shall happen to be.

17. The Committie of Estaites bothe at Edinburgh and the camp, considering that the instructiones heirtofore sent to the countrie for the use of the publict have been neglected and altogether slighted for the maist pairt, and the said Committie of Estaites finding thameselffes obliged be the trust and charge committed to thame, to provyde a tymeous remeid for preventing of such neglect and securitie in tyme cumeing, leist the not remeiding thairof indanger bothe the countrie and caus now in hand. Wherefore, they do heirby requyer all and everie one, in thair severall places and degries, to whom the obeying of thir instructiones are incumbent, that they exactlie fullfill and obey the above wrytten instructiones in everie poynt thairof, and mak speedie repoirt of thair dilligence thairanent, within the tymes above prescryvit; utherwayes, thir presents do certifie everie one who shall be deficient heirintill, that the next

instructiones, must and shall be militarie executione, of poynding be horss trowpers or foote companies, against those who shall be negligent, with libertie of frie quarters upon the delinquents, aye and untill they do thair duetie; and specially against the Committies of War, to whom the executing of publict orderes are principallie incumbent, and whose bygane neglect in thair places hath occassioned all the slighting of publict orderes throughout the whole countrie.

Act against Runawayes and Fugitives, and those who receipts, interteanes, conceals, and not apprehends or delates thame.

At Edinburgh, the eleveinth day of November, the yeir of God 1640.—The Lords and uthers of the Committie of the Estaites of Parliament, considering the fearfull prejudice and inconveience likely to ensew upon the frequent disbanding and steilling away of a number of base fugitives from thair companies and cullores. And withall, the said Committie of Estaites, considering that the maine and principal caus of the foirsaid disgracefull disbanding proceideth from the not examplare and exact punishment of the said base runawayes and thair receipters, and from the not putting of the former actes and constitutiones maid against runawayes and thair said receipters, and against idle and maisterless men, to dew executione. In consideratione whereof, the said Committie of Estaites doeth heirby ratifie and approve that former act maid against runawayes, maisterless men, and those who travelleth without testimonialls, of the date the xx. day of August last, bypast, in the whole poyntes thairof; requyering the samen to be put to dew executione by these to whom the executing of the samen is intrusted and committed, as they will answer upon thair heist paine.

And further the said Committie of Estaites, being now more sensible of the fearfull prejudice which the neglect of the executing of the said former act, conforme to the tenor thairof, hath produced, whereby the said base runawayes hath been enboldened to disband in frequent numbers, and are resaved and intertainit throw all pairtes of the countrie by disaffectit people, without anie kynd of punishment.—For remeid whereof, and without prejudice of the said former act, except in sua far as the samen is heirby renovat and reformed, the said Committie of Estaites doeth heirby straitlie requyer and command the whole Committies of War, Noblemen, Barones, Collonells, Gentillmen, Shereffes, Magistrates of Burrowes, Elderes and Constables, in each paroche, as they will be answerable to the said Estaites of this kingdome, that they and everie ane of thame do forthwith, immediatelie, caus mak present strait and exact search and tryall in all places where the said runawayes and fugitives, aither horss or foote, shall be found, and to delate, tak, apprehend and present thame befoire the Committies of War in each division, or Shereffes of the schyres, or Magistrates of the said burrowes, where the said fugitives shall be apprehendit, and whilk Committies, Shereffes and Magistrates shall be obliged to decimate and to hang the tenth man of thame; and if thair be bot one or more of thame within ten, to caus hang one of the said number, albeit thair be bot one, conforme to the said former act; and to send the rest to the said Committie of Estaites, at Edinburgh, upon the expense of the publict, to be punished with a mark of infamie and sent bak to thair companies.

And because the receipters and interteaners of the said fugitives are no less worthy of punishment nor the said fugitives thameselffes, thairfore it is heirby enacted, statuted and ordainit that whosoever shall keip, resave or interteane anie fugitive, horss or foote, and shall not delate or delyver thame in manner foirsaid, the said receipters shall be repute and esteimed as disaffected and enemeis to thair countrie and caus now in hand, and shall be accordinglie punished be the said Committie of Estaites, or Committies of War where they dwell.

And for the better discoverie of the said receipters, and finding out of the said runawayes, and that they may have no starting hole nor resting place within this kingdome; the said Committie of Estaites doeth heirby statute, enact and ordain, that if anie persone within this kingdome, of whatsoever qualitie or degrie, efter publication of this present act, shall presume to keip, harbour or interteane anie fugitive or runaway, or shall have anie knawledge or intellegence where anie of the said fugitives or runawayes are or shall be, and shall not immediatelie aither delate or delyver thame, in manner foirsaid; the said personse, interteaners, keipers and conceallers of the said fugitives as is above prescryved, shall, *ipso facto*, forfault and lose the half of all thair moveable gudes; the ane half whereof shall be applyed to the use of the publict, and the uther half of the samen shall perteine to whatsoever person shall delate or qualifie the said interteanment, keiping or concealing the said runawayes as is above written; to whom the said Committie of Estaites promiseth heirby, upon thair honour and credit, to grant and dispone the present right and possessione of the samen. And further, the said persones, delaters of those who shall receipt, interteane or conceal the said fugitives, shall be thankfullie rewarded be the said Committie of Estaites, and shall be repute and esteimed gude and weill affected countriemen.

And becaus thair is a great number of all sortes of people latelie cume from the armie, and from thair quarters and companies within this kingdome, now on foote for the defence thairof, whereof sundries have obteinit pass upon assurance to return within a short space. Thairfore, it is statuted and ordainit, that whosoever shall not returne to thair cullores, within four dayes efter the publication heirof, at leist immediatelie efter the expyering of thair pass, shall be reputed and esteimit as fugitives, and shall be lyable and subject to thair censure and punishment foirsaid.

And if the Committies of War within each division

or anie person being thairupon, shall be negligent in conveining and taking order with the said runawayes, and thair receipters and conceallers, or shall be deficient in putting of this act, together with the said former act, to execution efter the forme and tenores of the samen, rexane; each persone of the said Committie of War shall be unlawed and fyned be the Committie of Estaites in the soume of thrie hundred punds Scots monie, for each failzie, toties quoties.—And if the Ministeres and Elderes shall be found deficient in delatting, and Captaines or Constables of parochess, or anie uther parochiner, shall be negligent in searching, apprehending and presenting of the said fugitives and maisterless men to the said Committie of War, or other Magistrates foirsaid, or in putting the said actes to due executione, so far as concerns thair pairt thairof, each one of thame who shall be found negligent shall be fyned be the Committie of War within thair bounds, or be the said Committie of Estaites, in the soume of ane hundred punds money; the one halfe of which fynes shall perteane to the publict, and the uther halfe to the pairtie delator of the said negligent persones, rexane in manner foirsaid.

And if it shall cum to the knawledge of anie person who hath or shall happen to outreik sodgers, horss or foote, that those outreiked be thame are disbanded and fled from thair cullores, the said outputters of thame shall be obliged to searche, seik and apprehend the said fugitives, throw the whole bounds of the presbiterie where the said outputters dwelleth, and shall aither apprehend thame or put thame from that bounds; or utherwayes, in caice of thair neglect to do thair exact dilligence thairin, the said outputters shall be obliged to mak up thair number, be outputting of thame in thair places, sufficientlie provided in armes and uther necessares, upon the said outreikers thair own expenses

And ordaines thir presents to be published at the mercat croces of all heid burrowes, and the whole paroche kirkes within this kingdome, that none pretend ignorance heirof.—Printed at Edinburgh by James Bryson, 1640.

Answeris to the first artickle of the last Instructiones.

The whilk day, for obedience of the first artickle of the instructiones sent be the Committie of Estaites to the Committie foirsaid, daitit the 16 November, 1640, anent runawayes. The Committie hes delyverit all the said actes, to the Captaines of the parochess and uthers, to be given to the Ministeres, to be proclaimet at the paroch kirkes. And ordaines the said actes to be put in executione be the said Captaines, under the paines thairin conteinit.

Answeris to the second artickle of the Instructiones.

The Committie foirsaid, taking to thair consideratione the actes maid anent fugitives and runawayes, for putting of the samen to dew executione, be thameselffes, thair accessores and uthers, and to tak strict tryall for all these runawayes, idle and maisterless men within thair division; ordaines the Captaine of ilk paroche to tak and apprehend all fugitives, runawayes, idle and maisterless men, within thair bounds, and bring thame and thair resetters to Clauchanpluk, the tent of this instant. And that the said Captaines tak notice and tryall that no man pass throw thair bounds, aither horss or foote, wanting ane pass. And if they be found wanting ane pass, to apprehend suche and bring thame to the Committie. And also, the said Captaines cause the minister and elderes to tak notice that all the men persones within thair parochess, have ane testificate where they served the last terme; and all that wants suche, to bring thame to the Committie. And that, the said Captaines put in executione the said actes maid anent maisterless men and runawayes; with certificatione to ilk Captain that failzies in the premises, efter tryal, the said actes shall be put to executione against thame in the artickle that concerns thair fynes.

Answeris to the third articKle of the Instructiones.

The Committie foirsaid have sent all the clothe they could get for money, alreddie to Edinburgh, and as for schoes thair is nane to be had, aither maid or unmaid, in this countrie.

Act for apprehending Bagbie's wyff.

The quhilk day, anent the vilepending of the Commissar Depute, his commissione be Margaret Dunbar, spouse to Gilbert Browne of Bagbie, and covenanter, in intrometing with, useing and disponeing upon the crop of cornes in the Nuntone, perteining the said Gilbert, and quhilk was ordainit to be intrometit with be the Commissar Depute for the use of the publict; ordaines Johne Fullartone of Carletone, Captaine of the parochen of Twyname, to tak and apprehend the said Margaret Dunbar and put her in suir ward, thairin to remaine, till she find cautione to abstein in tyme cumeing, in respect hir intromissione was cleirlie proven to the said Committie.

Act—Apryseris of Thrie merk-land's cornes and uthers.

The quhilk day, compeirit the persones underwrytten, (wha were appoyntit for apryseing of the crops and uthers efterspecifit, perteining to the persones underwrytten,) they are to say—Robert Conquhor in Balgreddane, Donald Willsone in Halkit Leathes, Johne M'Gill in Guffogland, and Roger Moresone in Cassillgour. And deponit, upon thair great oath, according to thair knawledge, that Johne Maxwell of Thrie merk-land has, within the barne and barne-zeard of Mylnetown, xx laids aites, four small bolls beir; and on the ground, twa ky worthe xl merks, ane naig xl merks.

Item,—George Maxwell, brother to the said Johne Maxwell has, in barne and barne-zeard, xxiiij laids aites; and upon the ground, ane meir x merks.

Item,—Johne Makcartney of Leathes has, within the barne and barne-zeard of Leathes, xxxvj laids aites and four bolls beir; and on the ground, thrie naigs xxiiij libs.

Item,—Johne Maxwell of Collignaw has in his barne, four laids of aites.

Item,—Johne Browne of Mollance has, within his barne and barne-zeard, xxviij laids aites and thrie small bolls beir.

Act contra Robert Makculloche.

The whilk day, anent the alledgance proponit be Robert Makculloche of Kirriclauche, that he was ever willing to have subscribed the generall band, and offerit to Erlistone to have subscryvit the samen, but thair was not a notar. Whilk alledgance Erlistone denyit. Thairfore, the Committie ordaines the said Robert to prove the said alledgance be anie twa gentillmen, the next Committie day, at Cullenoche, the tenth of this instant; with certificatione if he failzie in probatione, the said Robert to pey the haile fyne laid on him, and if he compeir not the said day, to pey the double of his fyne for his contumancie.

Act—Pryce of Victual.

The quhilk day, the Committie ordaines, that the pryce of the laid aites and boll beir of the non-covenanters' crops, be at fyve punds the laid aites, and the small boll beir fyve punds, for the crop 1640. This pryce to be peyit within a month, or else, at Whitsonday, to pey aught punds for the laid aites and boll beir, overheid, messor of Kirkcudbryt.

The Committie foirsaid, halden be a sufficient coram, at Kirkcudbryt, the secund day of December, 1640. Erlistone preses.

Answer to the first artickle of the Instructiones.

The quhilk day, the Committie foirsaid, taking to consideratione the charge laid upon thame be the late instructiones and uthers, for assisting and furthering of the Commissar Depute within the said bounds, for inbringing of the rents, feu dewteis, teinds, dewteis of lands, and uthers quhatsumever, due to His Majesty, pretendit bischopes, papistes, ante-covenanters, and uther unfreindes within the presbiterie; doeth ordaine, heirby, that ilk Captaine, within thair division, within the said presbiterie, caus summond be wryt, all and quhatsumever persones due to His Majesty, pretendit bischopes, and uthers above specified, anie feu dewteis, teinds, mails, fermes of lands, soumes of money, and uther gudes and geir quhatsumever, to compeir before the said Committie, at New-Galloway, the xvij of this instant, bringane with thame, all and sundrie, the said teinds, feu dewties, and uthers above exprest, due to His Majesty, pretendit bischopes, and uthers above conteinit; and that restane awane of all yeires and termes bygane, since thair last lawfull discharge; with certificatione if they failzie, conforme to the general instructiones, thair shall be men sent furthe in armes upon thame, with power of frie quarteres, aye and until they mak peyment, as is above exprest. And also doeth ordaine, that the said Captaines inbring the tenth and twentieth penny rent, and mak peyment thairof to the Collector appoyntit for resaveing the samen, and that betwixt and the said xxvij of this instant, at New-Galloway; with certificatione to ilk Captaine that shall failzie in the commissione, he shall be committed to ward, thairin to remaine, aye and until he pey the double of the said tenth and twentieth penny rent within his division. And ratifies all former actes maid anent the inbringing of the said tenth and twentieth penny as befoire, within the said presbiterie.

Act—Barcaple.

The whilk day, anent the declaratione proponit be David Arnott of Barcaple, Captaine of the parochen of Tongland, declareing that Robert Ashennane of Cullquha, and Johne Aschennane of Barnecrosche, hes no meanes to pey thair tenth and twentieth penny rent except thair crop. The Committie ordaines the said Captaine to cause thresche out als muche as will pey the said tenth and twentieth penny, at fyve pund the laid.

Act—Civil Effaires.

The quhilk day, the said Committie findes be the warrand sent to thame frae the Committie of Estaites, that they have power to sit upon civil effaires, and that all pairties who hes anie controversies betwixt thame shall upon lawfull pursute have justice.

Act—Commissar Depute.

The quhilk day, the said Committie ordaines William Glendonyng to intromet with, use and dispone Gilbert Browne of Bagbie, his crop in the Nuntone, to the use of the publict.

Underwrytten pryce of the victual advanced be Bagbie's tennents, meill and beir overheid, aucht merks

And ordaines, that Johne Robesone in Kirkeoche deteane in his awn hand, out of the xij bolls beir that he was due to the publict for the dewtie of the said lands, 1640, the soume of fiftie punds money, advancet be him for the tenth and twentieth penny rent of the lands of Nuntone, quhilk perteins to the publict and of befoire to Bagbie.

The Committie foirsaid, halden be a sufficient coram, at Kirkcudbryt, the third day of December, 1640. Collonell preses.

Act for aprysing Robert Maxwell and Harry Lyndsay's Cornes.

The quhilk day, the said Committie ordaines the persones following to be aprysers of the cornes underwrytten, viz,—Adam Wright in Dundrennan, Johne

Cultane and James Malcome, thair. Quha being sworne upon thair greit aithe, that Robert Maxwell of Culnachtrie and his tennents hes perteining thame, in barne and barne-zeard, fiftie sex laids aites and four bolls beir, whereof perteins to the said Robert twa pairts. Item,—viij oxen, at xx libs. the peice. Item,—thrie kye, at ten punds the peice. Item,—xx scheip, pryce xl libs.

And that Harry Lyndsay of Rascarrel hes perteining him, in the barne of Rascarrel, twentie aught laids aites and four bolls beir Item,—four oxen and twa ky, pryce foirsaid.

Act—Captaines of Parochess.

The quhilk day, the Committie ordaines, that, ilk Captaine, within ilk parochen, tak up a list of all men able to beir armes within thair parochen, and to give in the samen, subscryvit with thair handes, at New-Galloway, the xvij of this instant; and how each man is provydit.

The Committie sworne.

The quhilk day, the persones of the Committie underwrytten, viz,—Alexander Gordon of Erlistone, William Gordon of Shirmers, William Gordon of Kirkconnell, Johne Fullartone of Carletone, Thomas M'Clellane of Collyn, Lancelot Greir of Dalskearthe, George Glendonyng in Mochrum, David Arnot of Barcaple, William Glendonyng, late provest of Kirkcudbryt, Alexander Gordon of Knockgrey, Alexander

Gordon of Garlarge, Robert Gordon of Knockbrex, Johne Ewart, baillie of Kirkcudbryt, William Lyndsay in Fairgirthe, Hew Maxwell in Mersheid, Robert Gordon, brother germane to Johne Gordon of Cardyness, being admitted and sworne upon thair great oath, deponed, that they and ilk ane of thame, shall faithfullie doe and exercise all things incumbent to thame, as the Committie of War within the Stewartrie of Kirkcudbright, for the weill and furtherance of the publict caus, and quhilk may advance the samen and to keip secreit.

The Clerk sworne.

The quhilk day, Robert Gordon, notar, Clerk to the said Committie, gave his oath *de fideli administratione* of his place and charge.

Act for Collyn.

The quhilk day, anent the supplicatione presentit be Thomas M'Clellane of Collyn, declareing the inhabilitie of his bodie to discharge the office of a captaine, within the parochen of Rerwick, quhilk was laid upon him; taking the samen to thair consideratione, findes the said Thomas M'Clellane unable for that charge, in respect of his age and seikness, and thairfore liberate him thairof. And ordaines Johne Cutlar of Orroland and Johne M'Clellane of Auchenguill, conjunctlie, to discharge and supplie that place, in all tyme heirefter, as they will be answerable upon thair perrall.

Act—Captaines Johne Gordones for thair pey.

The quhilk day, anent the supplicatione presented be Captaines Johne Gordon of Cardyness and Johne Gordone in Rusco, for thameselffes, and | in name of Captaine James Gordon, Captaine Lieutenant Forrester, and in name of thair officers and souldiers, schawing that they want unpeyit to thame and thair souldiers the third pairt of thair fourtie dayes lone, quhilk should have been peyit to thame furthe of the tenth penny as uther regiments were, desyreing the samen to be peyit furthe of the said tenth or twentieth penny rent; taking the samen to thair consideratione, findes the desyer of the said supplicatione most reasonable. And, in speciall, seing that Thomas Lord Kirkcudbryt, thair collonell, purchasit ane warrand for uplifting of the said tenth penny, in manner and for the caus foirsaid, of the date the tenth of Julij, 1640, doeth ordaine William Griersone of Bargaltone, collector of the said tenth and twentieh penny, to mak peyment to the said captaines, thair officers and souldiers, of the third pairt of the said fourtie dayes lone, conforme to thair accounts, to be sein and approven be Johne Fullartone of Carletone, the said William Griersone, William Glendonyng Commissar Depute, and Robert Gordon, Clerk.—Quhilk accounts, with thair discharge and ane extract heirof, shall be ane warrand to the said collector for the foirsaid peyment.

Act—Robert Ewart.

The quhilk day, the said Committie ordaines that the victual adebtit be Robert Ewart, burges of Kirkcudbright, to the publict, be at ten merks the boll.

Act—Erlistone and others.

The quhilk day, anent the supplicatione presented be Alexander Gordone of Erliston, schawing that more than ane yeir since having, for the Stewartrie of Kirkcudbryt, advancet to the publict the number of fourtie sex rex dollares, quhairof resavit ten dollares and thrie allowit for himself, and rests to him threttie thrie rex dollares; desyreing that he might be peyit aff the publict for his rests, or sum uther cours taken for his peyment; taking the samen to thair consideratione, findes that the samen, with sex rex dollares advanced be William Gordone of Kirkconnell, and four rex dollares advanced by David Arnot of Barcaple, should have bein collected and inbrought from the gentrie and heritores of the Stewartrie. Ordaines that the samen be peyit be the said heritores and gentrie, with four score punds awing to young Erlistone the time he was Commissioner at Edinburgh for the Stewartrie, to be imposit upon ilk persone conforme to thair rent. And, in the mean tyme, untill the samen be inbrought, that the said soumes be advanced be the Commissar Depute aff the foir end of the fynes; and that the gentrie and heritores repey the said Commissar againe, according as the samen shall be imposit upon thame.

Act—Minister of Tongland.

The quhilk day, the Committie ordaines Mr George Rutherford, Minister at Tongland,[1] to have the tymber maid worke, kysts, arks, and uthers, that perteinit to Mr James Scott, ante-covenanter, for peyment.

Act—James Lidderdaill, fuer of Isle.

The quhilk day, the Committie, taking to thair consideratione the supplicatione given in to thame be James Lidderdaill, fuer of St. Marie's Isle, schawing that he hes warrand and commissione from the Committie of Estaites for intromission with the cornes, rents and uther gudes, quhilk perteinit of befoire to Thomas Lidderdaill of St Marie's Isle, his father, to the use of the publict, as the said commissione producit proports, and that the said Thomas Lidderdaill vilepends the foirsaid commissione, be intrometing with the cornes, crops and other gudes and geir being within the said St Marie's Isle. And efter tryall taken be the said Committie of the said Thomas his intromissiones, be probatione of divers witnesses, Thairfore, doeth ordaine James Tailfeir of Haircleuche, Captaine of the paroche of Galtway, immediatelie efter the sicht heirof, to pass, searche, seik, tak and apprehend the said Thomas Lidderdaill, wherever he can be apprehended within the bounds of the presbiterie, and if neid beis to mak open dores and yetts, and to put him in sure ward, firmance and

[1] See Appendix.

captivitie, aye and untill he find sufficient cautione to absteine from intromissione with the said cropes and uthers above conteinit. And willes the said James Lidderdaill, to intromet with, use and dispone upon all and sundrie the foirsaid crop, corne, catell, insight, plenishing and uthers, quhilk perteinit to the said Thomas, his father, and now to the publict. And desyres, heirby, the provest and baillies of Kirkcudbryt to delyver to the said James Lidderdaill certane insight and plenishing, quhilk his said father hes in ane hows perteining to him, within the said burgh, and gif neid beis to mak open doores. And ordaines all gentillmen and uthers within the paroche of Kirkcudbryt, to go with the said James Tailfeir, for assisting him in putting of the commissione to executione.

Act—Mr Johne Corsane.

The quhilk day compeirit Mr Johne Corsane,[1] provest of Drumfries, in presence of the Committie, and produced, to be registered in thair buikes, the

[1] Mr John Corsane, advocate, was a considerable time provost of Dumfries during the Civil Wars. He was a lineal descendant of Sir James Corsane, a cadet of the family of Glen, who settled at Dumfries, where he increased in riches and honour, and had a lineal succession of heirs-male for eighteen generations, all of whom bore the name of John. Mr John Corsane was married to Margaret Maxwell, one of the daughters and coheiresses of R. Maxwell of Dinwoody, by whom he got the lands of Bardennoch; and, afterwards when he had put his son John in possession of Meikleknox, he sometimes designed himself Mr John Corsane of Bardennoch. Besides his country estates in Nithsdale and Galloway, it is said that he had a third part of the Burgh of Dumfries and lands belonging thereto, indeed many of the old houses there still bear the arms of the family.

commissione underwrytten; quhilk they did allow and ordaine the samen to be registered; whereof the tenor followes.

These are to give full power commissione and warrand to Mr Johne Corsane, provest of Drumfries, to resaive from the Commissares or Collectores of the tenth and twentieth pennies and rentes of unfreinds and bischopes within Galloway, all suche soumes of money as they have in rediness for the use of the south regement; with power to him to give acceptances and discharges of his receipt thairof, quhilk shall be as valid and sufficient to the foirsaid Collectores as I had given thame discharges myself; and whereanent I obleis me to renew thame discharges myselfe, upon sight of the provest's discharge, be thir presents wrytten be Mr Cuthbert Cunynghame and subscribed with my hand, at Drumfries, the last November, J$^{m.}$ VI$^{c.}$ and fourtie yeires, befoire thir witnesses, Roger Kirkpatrick, baillie of Drumfries, and the said Mr Cuthbert Cunynghame.

(sic subscribitur,) Home.

Roger Kirkpatrick, witnes.
Cuthbert Cunyngham, witnes.

Lykeas, I, James Hairstanes, Commissar and Collector appoyntit within the shereffdome of Nythisdaill, allowes and approves this warrand and commissione above wrytten, in all the heids and artickles thairof, be thir presents, subscribed with my hand, day, monethe, and yeir of God, and befoire the witnesses above specefeit.

(sic subscribitur,) James Hairstanes.

Roger Kirkpatrick, witness.

Cuthbert Cunyngham, witnes.

The Committie foirsaid, halden be a sufficient coram, at Cullenoche, the tenth day of December, 1640. William Griersone of Bargaltone, preses.

The quhilk day, the Committie ordaines, that, all the runawayes aprehendit, or to be aprehendit, be incarcerat within the tolbuthe of Kirkcudbryt, aye and whill they aither be sent to the Committie of Estaites or sent to the armie, and that upon the expenss of the publict, except these that finds suretie to be answerable, upon advertisement, to be reddie to march. And also, that maisterless men aprehendit, or to be aprehendit, be sustenit and outreiked upon the expenss of the publict.

Act—Barley and uthers.

The quhilk day, the Committie ordaines that the possessores of the lands of Barley, Corscraig and Barneschalloche, amongst thame, outreik ane foote souldier, and to have him at the Newtown of Galloway upon the xvij of this instant, in redienes.—And that Schirmers nominate anie persone he thinks fit in anie of the said lands.

Act—Dalskarthe.

The quhilk day, the Committie ordaines Lancelot Greir of Dalskarthe to delyver to Captain Gordone ane man for Starrieheuch.

Act—Margaret Dumbar.

The quhilk day, the Committie taking to thair consideratione the supplicatione presented in name of Margaret Dumbar, spows to Gilbert Browne of Bagbie, desyering that ane mantenance may be allowit to hir and hir childrene for thair mantenance and sustentatione. Efter consideratione thairof, ordaines the supplicant to meane herselfe to the Committie of Estaites. And whill the first of January next to cum, unto the tyme she use hir dilligence with the said Estaites, ordaines William Glendonyng, Commissar Depute within the presbiterie, to give to the said supplicant thrie small bolls meill, whereanent this present act shall be his warrand.

Act—Trublers of uthers.

The quhilk day, the Committie finding that quarrells and jarrs amongst certane persones, are an impediment in stoping of thame to cume to thair meitings for discharging thair employments and doing thair necessar effaires. Ordaines that if anie persone, of quhatsumever degrie or qualitie they be of, shall upon reviving of anie former quarrell or uther accidents, happen to truble or molest, in bodie or gudes, anie thair adversaries or uthers, the tyme that they are at the meitings or in cumeing or goeing thairfrae, that persone or persones, sua committane wrong, according to thair degrie and to the degrie and rank of the persone harmet and to the wrong done, shall be condignelie punishit and censurit in bodie and gudes.

Act—Apprysers of Partone's cornes.

The quhilk day, the persones underwrytten, appoyntit for appryseing the lard of Partone's crope, they are to say, William Martene of Dullarg,[1] William Gordon in Craichie, Alexander Makill in Arnemannoch, and Fergus Neilsone in Glenlair, compeirit; and being solemnlie sworne upon thair great oathes, depones, that according to thair knawledge, the said lard of Partone not to have above xx laids aites and ane laid beir in the barne and barne-zeard of the Boreland of Partone.

---

[1] James Martin of Dullarg suffered severely during the persecution, both from fines imposed and from soldiers being quartered upon him. At length in 1664, having at the instigation of Mr Colin Dalgleish, curate, been fined in a thousand pounds for his wife not keeping the church, he was thrown into prison until the fine should be paid, where, owing to the severity used and the want of accommodation his health was so injured that he died in a short time. His son William Martin also sustained much loss from fines and exactions at that time.

Letter in favores of Johne Newall.

Ryght Honourabill,—It is represented to us be Johne Newall, that you have fyned him in fyve hundred merks for not subscryveing of the generall band. We resolve to give no answer to his supplicatione till we be informed be you, aither be wryt or be ane of your number, of the grounds and reasones wherefore ye have fyned him; whereof we desyer to be informed with all convenient dilligence; and in the meantyme you shall spair the exacting of the said fyne till the bussiness shall be considderit heir.—And so we rest your affectionate freinds.

When ye inform us aither be wryt or be ane of your number of the premises, ye shall adverteise Johne Newall, to the effect he may be present heir to answer for himselfe.

(sic subscribitur,) Cowpar.
Craighall.
Caprington.
Murray.
Dundas,
Edward Edgar.
Richard Maxwell.
Mr Wm. Moir.

Edinr., 2 Dec., 1640.

To the Ryght Hon. the Lord Kirkcudbryt and Committie of War thair.

The Committie halden be a sufficient coram, at New-Galloway, the xvij day of December, 1640. Erlistone preses.

Letter anent the third part of the Regiment at Drumfries.

Ryght Honourabill,—The inhabitants of the burghe of Drumfries have represented to us thair hard estate and conditione in billating of the southe regiment, for the quhilk thair poore people getts not tymelie satisfactione, and quhilk they cannot endure unless a cours be taken for thair releiff and peyment.

It is not unknawn to your lordship that, among the rest of the dews of that countrie, we have alloted all that is dew to the publict in your lordship's divisione for mantenance of that regiment; and notwithstanding of the severall orderes sent from this to hasten the uplifting and peyment of what is dew within your bounds to the Commissar, for the effect foirsaid, yet we find nothing effectuate nor performed, quhilk we conceive to proceid from the unwillingness of the countrie people or the slackness of the Commissares or Collectores in your divisione, or from the skarcietie of money. Quhairfore, and for your lordship's ease, and that the sogores may live upon suche commodities as that countrie affords, we have resolvit that the regiment shall be devydit in thrie third parts—quhairof, ane to lye at Drumfries, ane uther within your lordship's divisione, and the third within Lord Johnstone's divisione—for we find it impossible to the towne of Drumfries to interteine the regiment unless they be suppliet with the dewes of your lordship and the Lord Johnstone's divisione; and your people will be more able to billet thame with thameselfes nor to lift soumes of money for that effect.—Yet befoire we give orderes heiranent, we have resolved to acquaint your lordship, thairwith, intreating your lordship's assistance, that if they be sent thair your lordship wald be carefull of thair mantenance. Bot if the regiment could be keipit togither we wald rather wish it, quhilk cannot be unless your lordship caus hasten the uplifting and peyment of all that is dew within your divisione, suche as—the tenth and twentieth penny, ante-covenanter's and papist's rents, and uther dews to the publict, conforme to the generall instructiones, and cause the samen to be delyverit to the Commissar at Drumfries for the use of said regiment. If the Commissares and Collectores thair do not thair duetie to hasten in thir dews, we will be necessitat to give orderes to the Commissar of the regiment to uplift what is dew.

Your lordship hes to remember that you are obleist as Collonell in that divisione, to report exact dilligence and answer to the printed instructiones within the tyme limited thairby. Quhairfore these are to remember that the samen be not neglected, and specially that report be maid of all the numbers of horss and foote with armes dew furthe of your divisione, and that they be in reddiness upon advertisement, as weill the 4th and 8th man and trowper horss for the first leavie as the 10th man and trowper horss for the receive.

Thus being confident of your lordship's care and performance of thir particulars, we continew your lordship's affectionate freinds.

Postscript—If money cum not in to the Commissar for the use of the regiment befoire the xxth of this instant they cannot indure longer delay, and they have orders to devyde efter that tyme in caice betwixt and that they get not a supplie.

(sic subscribitur,) Argyle.
Burghly.
Cowpar.
Murray.
Craighall.
Hamilton.
Mr Wm. Moir.
Thos. Patersone.
Edward Edgar.

Edinr., the 10th Decr., 1840.

To our Noble Lord the Lord Kirkcudbright.

Ane Cold Covenanter.

The quhilk day, the Committie foirsaid finds and declares ane cold covenanter to be suche ane persone quha does not his dewtie in everie thing committed to his charge, thankfullie and willinglie, without compulsion, for the furtherance of the publict.

The quhilk day, Alexander Gordon of Knockgrey, Captain of the parochen of Carsfarne, declares no cold or uncovenanters within that parochen.

Alexander Gordon of Erlistone declares no cold or uncovenanters to be within the parochen of Dalry, whereof he is Captain, except Johne Newall

Alexander Gordon of Gairlarg, Captain of the parochen of Kelles, declares no cold or uncovenanters to be within the said parochen of Kelles.

William Gordon of Shirmers, Captain of the parochen of Balmaclellan, declares no cold or uncovenanters within his parochen.

George Glendonyng of Mochrum, Captain of the parochen of Partone, declares the lyke.

George Livingstone, Captain of the parochen of Balmaghie, declares the lyke.

William Gordon of Kirkconnell, Captain of the parochess of Buittle, Crocemichael and uthers, declares no cold or uncovenanters within his bounds except John Maxwell of Mylnetone; William Maxwell of Midkeltone; Gilbert Maxwell of Slognaw; Mr Patrik Adamsone, sumtyme Minister at Buttle; Mr James Scott, sumtyme Minister at Tungland; George Tait; Paul Reddik; Johne Browne of Mollance; Robert Browne, his brother; Johne Maxwell of Colignaw; James Maxwell of Brekansyde; Thomas M'Gill at Keltone.

Willam Lindsay, Captain of the paroches of Colvend and Suthik, declares no cold or non-covenanters within these parochess, except James Lindsay of Auchenskeoch; Andro Lindsay, his sone; Robert Lindsay, his sone; Charles Lindsay, his oy; Johne Lindsay of Wachope; Charles Lindsay, his uncle; Lancelot Lindsay, brother naturall to Wachope; Johne Lindsay, his brother naturall; Roger Lindsay of Maynes; Johne and James Lindsayes, his sones; Charles Murray of Barnhurrie; David Lindsay, sone to James Lindsay of Fairgirthe; Richard and William Herreiss, brethern to Robert Herreis of Barnebaroche; and the said Robert, late covenanter.

Robert Maxwell of Cavens, Captain of the parochen of Lochruttone, declares no cold or uncovenanters within his bounds, except Edward Maxwell of Hills; William Maxwell, his sone; Alexander Maxwell, his naturall sone; Lady Auchenfranko; Richard Herreis, hir sone; Edward Maxwell, callit of Carswada; Gudewyfe of Hills; and Johne Welshe in Langwodheid.

Johne Cutlar of Orroland, Captain of the parochen of Rerrik, declares no cold or uncovenanters within his bounds, except Robert Maxwell of Orchardtone; William Makclellane of Airds; William Makclellane of Overlaw; Robert Maxwell of Culnachtrie; Harie Lindsay of Roscarrell; John Makclellane of Gregorie; William Makclellane of Meifeild.

Lancelot Grier of Dalskarthe, Captain of the parochen of Troqueer, declares no cold or uncovenanters within his bounds, except Johne Maxwell of Kirkconnell; Elizabeth Maxwell, his mother; Helene Maxwell, Lady Mabie; Johne Herreis of Mabie, hir sone.

Captain of the parochen of Newabbay declares no cold or uncovenanted within his bounds, except James Maxwell of Littlebar.

Captain of the parochen of Kirkbeane declares the lyke, except Johne Sturgeon of Torrerrie; Johne Sturgeon in Cowcorse.

James Smithe, Captain of the parochen of Irongray, declares no non-covenanters within his bounds.

Johne Reddik of Dalbeattie, Captain of the parochen of Urr, declares the lyke.

Roger Maknacht of Killquhenatie, Captain of the parochen of Kirkpatrick, declares the lyke.

Johne Fullartone of Carletone declares the lyke.

David Arnot of Barcaple declares the lyke.

Richard Muir of Cassincarrie declares the lyke.[1]

Act—Commissares

The quhilk day, the Committie taking to thair consideratione the artickle of the instructiones anent the assisting of the Commissares and Collectores, for ingetting of what is dew to the publict, and of the severall actes maid by thame thairanent, and that the samen are slighted be the persones debtores,

1 For Extracts from Dumfries Session Book regarding non-covenanters, mentioned in this list, see Appendix.

doeth of new againe ordain, that all persones of whatsumever qualitie and degrie they be of, who are adebtit any soumes of money to the publict, that they mak full and complete peyment thairof to the Commissares and Collectores, betwixt the date heirof and the last of this instant; and that aither in money, or meill or beir at v lib the boll, or in ky or oxen, the ox at x lib the peice and the kow at ten merks, to be delyverit to the Commissar in Kirkcudbryt betwixt and the said day; with certificatione to thame that failzies, that men in armes shall be sent furthe upon thame upon frie quarters aye and untill they pey the samen. And ordaines that ilk paroche send ane musqueteir to Kirkcudbryt, upon the last of this instant, to the effect foirsaid. The Commissar to be commander and Johne Howie his lieutenant.

Act contra Non-Covenanters.

The quhilk day, the Committie foirsaid ordaines, that the rents and estaites of non-covenanters, the awners quhairof are outwith the kingdome, be uplifted be the Commissar to the use of the publict.

Act—Robsone.

The quhilk day, the Committie foirsaid dismiss Thomas Robsone, runaway, and ordaines the parochen of Newabbay to put furthe ane uther in his stead.

The Committie foirsaid, halden be a sufficient coram, at New-Galloway, the xviij day of December, 1640. Erlistone preses.

The quhilk day, the Committie foirsaid ordaines Johne Hutchisone, wright, burgess of Kirkcudbryt, to delyver to the Commissar xl merks money, for the customes of the towne of Kirkcudbryt that he uplifted betwixt the first of November, 1639, and first of November, 1640; and to deteane the rest of xxxij lib in his awn hand, quhilk is over and abune the said xl merks.

Act—Johne Browne.

The quhilk day, the Committie foirsaid finds the factorie granted be Gilbert Browne of Bagbie to Johne Browne, merchand, (being eftir intimatione no to meddle with non-covenanters,) of the mertinmes term rent last of Bagbie, to be null, and he to be censurit for medling with the said Bagbie.

Act—Johne Browne.

The quhilk day, the Committie foirsaid finds that Johne Browne, merchand, be his letter producit, is tacksman of the lands of Camret for this yeir, 1640. And the Commissar to deduce to him xvj libs money, viz.—viij libs at Mertenemes and viij libs at Whitsonday, in respect it was about Lambmes befoire the said Johne enterit to the said lands.

Act—Johne Grier.

The quhilk day, Johne Grier in Larg, being convenit befoire the said Committie be the Commissar, for the dewtie of the lands of Larg, perteining to Bagbie, for the Mertenemes terme, 1638, deponed that the said dewtie is all peyit except foure merks money. And thairfore liberates him of the rest.

Act—Aprysers of Non-Covenanters Cropes.

The quhilk day, the said Committie taking to thair consideratione that the former orderes given be thame for apryseing of the non-covenanteres' cornes and gudes, aither out of ignorance and prevalitie hes been slichted and not truelie apryset, doeth now ordaine that the samen be apryset be the persones underwrytten, quha shall be sworne to apryse the said cornes and gudes to the utmost according to thair knawledge. They are to say—William Clinton in Carleton, Johne Schaw, Johne M'Dowell in Barholme, and Thomas Robsone, maltman, in Kirkcudbryt.

Act contra Captaines.

The quhilk day, the Committie ordaines that if anie Captain within thair divisione conceals anie rents or gudes perteining to non-covenanters or to the publict within thair bounds, that the concealler shall pay as much of his gudes or rent as he conceals, attour the rent or gudes conceallit.

Act in favores of the Commissar.

The quhilk day, the Committie ordaines that the Captaines of the parochess give command to all persones whatsumever that hes testaments to confirm within thair divisione, that they go to Kirkcudbryt, the last of this instant, and confirm the samen, under the paine of censure.

Act—Mr Hew Hendersone.

The quhilk day, anent the supplicatione presentit be Mr Hew Hendersone, Minister at Dalry, schawing that he hes servit the cure at the said kirk ane yeir and ane halfe bygane without anie peyment of steipand; and that the parochinares promesit to have buildit him ane hows; and that the said parochinares, at the leist the maist part of thame, had condiscendit for building of the said hows, to pey to him aught merks of the merkland, for the quhilk ane great part had given band. Notwithstanding, for the present, the said parochinares will on nawayes mak peyment of the said steipand or for the building of the said hows, as the said supplicatione beires. The quhilk supplicatione being heard, scin and considerit, doeth decerne and ordaine that the parochinares of the said parochen and uthers who are adebtit and in use of peyment of the teinds of the said parochen, redelie answer intend, obey and mak thankfull peyment to the said Mr Hew, of the teinds of the lands, conforme to use and wont, and that restane awane of all yeires and termes bygane since his right to uplift the samen. And also ordaines that the said persones mak peyment to the said Mr Hew of the said soume of aught merks furth of ilk merkland, in manner and for the causes foirsaid and as is conteinit in the said band. And ordaines the Captain of the paroche to sie the said Mr Hew compleitlie peyit, conforme to the tenor heirof, and if neid beis to poynd and distrenzie thairfore, ilk ane for thair awn pairts, the ox ten punds and the kow ten merks, conforme to the generall order.

Act—James Gordon.

The quhilk day, anent the supplicatione presentit be James Gordon of Croftis, schawing that he hes money to pey to the publict and that the persones who hes the samen to lend upon suiretie will not without ane warrand to that effect, as the said supplication beirs. The quhilk being heard and considerit be the said Committie, finds the desyer of the samen most reasonable, and thairfore grants the desyer thairof, and dispenss with anie persone whatsumever to lend him money upon suirtie.

The Instructiones answerit.

The quhilk day, the printed instructiones sent to the Committie, the 16 of November, 1640, be the Estaites, is answerit in manner following.

1. Our fugitives are brought in and shall be delyverit according to the orderes given.

2. Thair is cours tane be us for preservatione of all hieghwayes, for apprehending of runawayes and maisterless men.

3. The clothes which the countrie can afford is alredie sent out, and as for schoes the countrie affords thame not, as we did formerlie demonstrate.

4. Ressave heirwith the roll of ante-covenanters or late covenanters. As also, we have given in ane trew inventar and roll of all moneys dew to the non-covenanters and of thair rents to the Commissar.

5. We have taken order for assisting of the Collectores and Commissar for ingetting of what is dew to the publict.

6. We have advertisit our Collector and Commissar to repair to Edinburgh, for fitting thair compts.

7. Our valuationes was closet and sent to you in Maiche last.

8. Answered in the sixth artickle.

9. Our ministrie is going on with all dilligence in getting of voluntar contributiones.

We have in September last sent furthe our silver plate, and what is not sent shall be sent furthe.

11. It shall be answerit neir the tyme prescryvit.

12. Our levies of foote was perfyted to the full.—Our horss, that are inlacking onlie nyne of our number, against the persones that put thame not out Erlistone sent out the executiones, particularlie in August last.

13. We never resaivit orderes for that effect.

14. Our volunteires shall be redie upon adverteisment.

15. Ressaive heirwith ane roll of the names of our Committie and Clerk.

Certaine of the Committie sworne.

The quhilk day, the persones underwrytten of the Committie of the Stewartrie compeirit and deponit as is befoire conteinit and as the rest depouit, and that upon thair great oathes: they are to say,— Johne Vyscount of Kenmore; Johne Lennox, elder of Callie, and in his absence Alexander Lennox, his sone; Roger Maknacht of Kilquhennatie; Robert Maxwell of Cavence; Richard Muir of Cassincarrie; Johne Cutlar of Orroland; and John Reddik of Dalbeattie.

Process contra Johne Newall.

Informatione to the Committie of Estaites frae the Committie of War appoyntit within the Stewartrie of Kirkcudbryt, contra Johne Newall.

1. The said Johne Newall was lang averse frae the subscryving of the covenant of this kingdome, as did apeir be being ane of the last subscryvers in the congregatione, albeit that he was bothe privatlie and publiclie delt with for that effect.

2. The said Johne altogither refuissit to subscryve the Declaratione of the Assemblie at Glasgow, in presence of the whole congregatione, to the evill example of uthers.

3. The said Johne, being baillie to the Erle of Galloway in that paroche, he never concurred in nothing that concernes the publict, bot be the contrair doeth controll the parochinares proceidings, in verbo at facto, in sua far as he hes power, to the evill example of uthers. Instance, his bitter railling contra Erlistone for overvalueing (as he alledges) of the said Erle's lands within that paroche, in presence of the minister, elderes, and the most considereble men of the paroche.

4. The said Johne is knawn to be ane ordinar murmurer and complainer contraire the proceidings of the Estaites and uthers, inferior judicatories for the tyme, for thair alledgit greivances and burdens, exactiones and impositiones prest upon the countrie.

5. The said Johne publictlie refuissit to subscryve the generall band, being requyered thairto be Erlistone, (to whom the oversicht of the subscryveing of the said band was committed,) to the bad example of uthers; quhilk sume hes confessit since, that his bitter curseing of himself that he should never do it, prevaillit upon thame and was the onlie ground of thair refuissall.

xviij December, 1640

The quhilk day, Alexander Gordon of Erlistone declares, that, conforme to the command given to him, he citat the said Johne Newall to compeir befoire the Committie, to heir and sie the witnesses underwrytten admitted and sworne befoire famous witnesses.

The said Johne being callit, not compeirand; admits for probatione, Robert Cannan in Blackmark, ane of the elderes of Dalry; George Logan in Boig; Johne Edgar in Ardoche; Johne Gordon in Glenhoule; and Johne Makcome in Stronepatrick; all elderes of the said paroche of Dalry; being all deiplie sworne upon the poyntes foirsaid, that ilk ane of thame shall the right suithe say and nae suithe conceal, according to thair knawledge.

Robert Cannan proves the first, secund, third and last poyntes.

George Logan, sic eque.

Johne Edgar proves the first four.

Johne Gordon, sic eque.

Johne Makcome proves the first, secund and third poyntes.

At New-Galloway, xviij Dec., 1640.

The quhilk day, the within named Johne Newall of Barskeoche, at command of the Committie of the Stewartrie of Kirkcudbright, being citat to have compeirit befoire thame, the said day, to answer to the indyctment within exprest He being callit and not compeirane, in manner and to the effect within specified, and divers famous witnesses, the elderes of the paroche quhair he dwells being citate, at command foirsaid, to beir witness to the said indyctment, they compeirit, and being deiplie sworne, sufficientlie did prove the haill indyctment, in everie poynt and artickle thairof, as is within rehearsit.

(sic subscribitur,) The maist pairt of the Committie.

The Committie foirsaid, halden be a sufficient coram, at Drumfries, the xxix day of December, 1640. Erlistone preses.

Act—Runawayes.

The quhilk day, the Committie ordaines that the minister and elderes of that paroche where anie runaway is visited with seikness,[1] testifie the samen under thair hands, upon thair consciences, of thair inhabilitie, utherwayes that paroche to supplie thair place.

Act—Johne Wilsone.

The quhilk day, the Committie finding that Johne Wilsone, runaway, furthe of the paroche of Lochruttone, is unable to doe service, thairfore liberates him and his resetters of censure.

Act—Johne Wilsone.

The quhilk day, the Committie finding that Johne Wilsone, runaway, in Crocemichael, is unable to goe upon service, thairfore dismiss him, and ordaines that he and his resetters, in all tyme heireftir, be frie of censure; and ordaines the said paroche of Crocemichael to furneis ane uther in his place.

---

1 For Notices regarding the Pest or Seikness see Appendix.

Act—Roger Oliver.

The quhilk day, the Committie ordaines that Roger Oliver, baggage man of Ironegray, be answerable for the baggage horss thairof, and that his crop be furthcumane to that effect, and ordaines Thomas Rome, Captaine of the said paroche, to be answerable for the said crop.

Act—Lard of Cardyness.

The quhilk day, the Committie ordaines, that the baggage horss of the parochess of Partone and Balmaghie be delyverit to Johne Gordon of Cardyness, quhilk is yet restane; and that the gudeman of Erlistone, James Tailfeir of Haircleugh, and William Gordon of Shirmers, pey to him ane uther restane.

Act—William Ker.

The quhilk day, the Committie admits William Ker for Johne Thomsone, runaway, in Tarregles.

Act—Mantenance of Runawayes.

The quhilk day, the Committie taking to thair consideratione, anent the mantenance of certane runawayes, within thair divisione, from the armie, (upon his Excellencie's letter and warrand sent to the said Committie to the twa Captaines, Johne Gordon of Cardyness and Johne Gordon in Ruscow, to that effect,) now aprehendit, extending to aught score sextene persones. Finding, that be the Act of the

Committie of Estaites, of the date the xj November last, should be interteinit off the publict. Thairefoir the said Committie ordaines that all runawayes, sua to be aprehendit and sent bak to thair cullores, shall have peyit to each of thame vj pence a day, for the space of aught dayes, for thair mantenance to the army. And ordaines the lard of Haircleuche, (in the absence of the Collector,) off the first and rediest of the tenth and twentieth penny rent within his subdivisione, to pey the samen at the sight of young Erlistone, upon the Captaines' discharge, quhilk togither with this act shall be ane warrant to the said lard of Haircleuche and Collector.

Act—Johne Stewart of Schambellie.

The quhilk day, Johne Stewart of Schambellie being convenit befoire the said Committie, at the instance of William Glendonyng, Commissar Depute, for peyment to him of the few mailles, fermes, dewteis, and personege teindes of the Four-merk-land of land and Five-shilling-land of Howlat Close, perteining to the pretendit Bischope of Edinburghe, lyeane within the parochen of Newabbey and Stewartrie of Kirkcudbright, for the crop 1639 yeires for the use of the publict. The said Johne alledgit that he could not pey the samen to the said Commissar, as belonging to the publict and as ane pairt of the said pretendit bischope's rent, in respect that William Maxwell of Kirkhouse, quha is tacksman to the said pretendit bischope, hes his band for peyment thairof to him.

Notwithstanding whereof, the said Committie ordaines the said Johne Stewart to pey the said rentes and uthers foirsaids to the said Commissar Depute. And declares the said Johne to be liberatit of all bands grantit be him to the said William Maxwell for that effect.

Act—William Makclin.

The quhilk day, admits William Makclin in the place of James Gibsone, runaway.

Act contra Dalbeattie and uthers.

The quhilk day, the Committie finding that severall of the Captaines of the parochess have been negligent of the charge committed to thame, and in speciall that of the inbringing of the runawayes. Thairfore, ordaines Johne Reddick of Dalbeattie, Captaine of the parochen of Urr, Johne M'Clellane of Auchengule and Johne Cutlar of Orroland, Captaines of the parochen of Rerrick, betwixt and the last of this instant, to inbring thair runawayes and delyver thame to the Captaines heir, at Drumfries; and for ilk man they failzie to produce to pey xl merks money, attour the production.

Letter from the Committie of Estaites for proclaiming the Actes anent Tanning of Leather, and uthers.

Ryght Honourabill.—Ressave heirwith a number of printed actes anent hydes, schoes, bootes, and tanning of leather. Ye must tak exact cours that thir actes be keipit, as weill to burgh as land, within your schyer, and they must be proclamit at your mercat croce and everie paroche kirke.

Ressave lykewise sum actes whilk was resolved upon with consent of the last members of the parliament, beirane that everie schyer and divisione shall provyde thair awn souldiors of clothes and schoes.—We do heirby most ernestlie recommend to you the prosecuting of this act, and that ye caus mak coppies thairof and send to all pairts of your schyer, and speciallie to the Committies of War; for we cannot express the miserie and hazard our poore souldiors are in to cover thair nakedness. It is pitie to sie the great securitie which is amongst us. Bot we are confident and we do verelie expect that ye will be carefull in this particular, and the conditiones of the act shall be reallie observit be you.

We doubt not bot you must consider what dilligence ye are tied to be the generall printed instructiones and actes against runawayes. Thair are sum of the dyetis of the report of your dilligence alreddie prescryved; bot be assured if we find not ane exact and particular report of your dilligence, with a reall performing of these instructiones, thair will be suche cours taken for obeying thairof as will be unpleasant. This we will not slip immediately efter the prescryving of the dyetis of the instructiones, and you may tak this as your last adverteisment.

You are also to ressave the doubill of an edict to warne all within your schyer to whom anie thing is restane, and the Collectores and Commissares, quhairof ye shall caus mak coppies and caus execute, conforme to the tenor thairof.

So to your dilligence and answer we rest your affectionate freindes.

(sic subscribitur,) Kinghorne.
Burghly.
Cowpar.
Craighall.
Murray.
J. Hamilton.
Edward Edgar.
Mr Wm. Moir.

Edinr., 17 Dec., 1640.

Act anent the Pryces of Schooes, Bootes, Hydes, and Tanning of Leather.—Nov. 26., 1640.

At Edinburgh, the twentie sexth day of November, 1640 yeires. The Committie of the Estaites of Parliament, with advyse of the Nobilitie, Commissioners of schyres and burrowes, convenit for the tyme, having appoynted a number of each Estaite to settle the pryces of the schoes, bootes, hydes, and tanning of leather. Which persones, efter full deliberatione thairanent, resolved and concludit upon these artickles underwrytten, which they presented and produced in presence of the said Committie of Estaites, Nobilitie, and Commissioners foirsaid. And efter the samen artickles were read and considderit, in thair presence, they all in ane voyce ratefeit and approvit the samen, as they are heirefter set down. Of the which artickles the tenor followeth.

These on the Committie appoynted for settling sum rules to the cordinares, have efter diliberatione considerit; that in respect the prycess of the rough hydes are almost at the samen rate, sold for the present, that they were sold for at this tyme twelff-month, have thocht thairfore expedient that the pryces of the rough hydes may be ordained to be sold in manner following, viz :—

That the best ox hyde be sold for viij merks, and inferior sorts of oxen hydes for v libs., vij merks, and iiij libs., and so furthe, according to thair worthe, being rough hydes.

Item,—The best kyne hydes, being rough, be sold for iiij libs, and the inferior sorts for v merks, iij libs., and iiij merks, and so furthe, according to thair worthe.

Item,—That thair be allowed for the tanning of the best ox hyde, for materials, paines and ganie, fiftie shillings; and for the secund sorts of ox hydes and kyne hydes, overheid, fourtie shillings.

And for sure performance heirof, it is ordainit, that, the Magistrates of each burghe, and Justice of Peace in landwart, shall have power to caus mak oppen the bark pottes, for visiting the leather, and for urging the tanners to sell the samen at the pryces foirsaid, at the discretione of the Magistrates and Justices of Peace, conforme to the order before prescryvit of the rough hydes. And the contraveiner, in refuiseing to sell the rough hydes at the pryces foirsaid, to pey xls. for each hyde; and the tanner, who refuiseth, to pey iij libs for each hyde bye and attour the fulfilling of the act; and the penaltie to be devydit as followeth:—the one halfe to the delaiter and the uther halfe to the Magistrates or Justices of Peace for the use of the publict.

And siclyke, ordaines that the cordinares sell the bootes and shoes as followeth, viz.:—

The inch of thrie-solled schoes, of the best leather, be sold at twa shillings twa pennies the inch.

Item,—The secund sort of thrie-solled schoes be sold at xxd. the inch.

Item,—The inch of single-solled schoes, of the best sort, at xvjd. the inch.

Item,—The secund sort of single-solled schoes, at xiiijd the inch.

Item,—the inch of barnes' schoes, double-solled, of the best sort, at xvjd.

Item,—The secund of slighter leather, double-solled, at xiiijd.

Item,—Of single-solled, at eight inches and under, at xijd. the inch.

Item,—That women's schoes, tymber heilled, of the best sort, be sold at xxvjd. the inch.

Item,—That the secund sort of tymber heilled, at xxd. the inch.

And anent the pryce of bootes, it is ordained, that thair be allowed of the best leather for each inch of the length of the foote of bootes the quadruple of the pryce of the inch of the best sort of schoes, being xxijd. for each inch thaiaof; the uppers being large and of the best leather.

Item,—That the secund sort of leather maid in bootes, that the inch of the length of the foote thairof be sold at the quadruple pryce of the schoes maid of secund leather, as is befoire prescryvit, extending the pryce of the said schoes of that sort to xxd.; the uppers thairof being also large.

And the said Committie of Estaites, with consent foirsaid, for sundrie causes and considerationes, have thocht fit for the burghe of Edinburgh, that the cordinares within the said burghe shall have for each inch of schoes of the best sort of leather xxxd., and for the secund sort of leather twa shillings for the inch, and so furthe in the lyke proportione in single-solled and women's shoes.

And the penalties against refuisers to sell at the pryces foirsaid, and the fynes for not sufficient geir to be confiscate, the one halfe to perteine to the delaiter, and the uther halfe to the judge, for the use of the publict. And who refuiseth to work and leiveth off, to pey fourtie punds, bye and attour punishment to be inflicted in thair persones, toties quoties, as they shall be challenged and fund guiltie, and thair penalties to be employed in manner befoire prescryvit, and these of the poorest sort to be punished at the discretione of the judge.

Whilk artickles above wrytten the said Committie of Estaites, with advyse of the Nobilitie, Commissioners of schyers and burrowes, ordaines to be enacted in the buikes of the Committie of Estaites. And ordaines the samen to be a rule to the whole kingdome, heirefter, and to be published at all the mercat croces of heid burrowes within this kingdome, and paroche kirkes, to the effect that none pretend ignorance thairof.—Printed at Edinburgh by James Bryson, 1640.

Act for Clothes and Schooes.

At Edinburgh, 26 Nov., 1640.—The Lords and uthers of the Committie of Estaites, be advyse of the Nobilitie, Commissioners of schyres and burrowes now convenit, haveing taken to thair consideratione the hard estaite and conditione that the souldiors of our armie are in for the present, for want of clothes, schooes and schirts. And notwithstanding of the frequent adverteisment and orderes sent from the said Committie of Estaites to the whole pairtes of this kingdome, requyering thame to provyde and furneis the said commodities, for the use foirsaid, and to pey the pryces of the samen out of anie thing dew be the countrie to the publict, yet thair is no considerable quantitie prepared and furneised for the said use; so that the souldiors will be in great danger unless a present remeid and cours be tane how they may be clothed and furneished with the said commodities, with the greatest expeditione. Quhairanent, it being much debated quhat may be the reddiest and most proffeitable cours to get the samen speidile effectuated, the said Committie of Estates, with consent of the said Nobilitie and Commissioners for schyers and burrowes, now presentlie convenit, have, be thir presents, fund and resolvit, statuit and ordainit, that everie schyer and divisione in the kingdome, as weill to burghe as landwart, shall with all expeditione furneis and provyde schooes and clothes for thair awn souldiors sent out from thair divisione; quhilk clothes and schooes shall be peyit out of the vouluntar contributione within thair awn bounds, and quhair the voluntar contributione is alreddie peyit, ordaines the said commodities to be peyit out of the tenth, twentieth penny, or anie thing else dew to the publict. And where it is lykewayes peyit and nothing restane to the publict, it is heirby declared, that the furneishers of the said clothes and schooes to thair awn souldiors, within ilk divisione, shall have securitie from the Committie of Estaites, in name of the Estaites of this kingdome, for the pryces of what they shall furneis and advance for the use foirsaid. And ordaines ilk schyer, divisione and presbiterie to mak present provisione, with all possible dilligence, for clothes and schooes to thair awn souldiors, sent furthe of thair divisione, and to send the samen to the Committie at Edinburgh or to the camp, for the use of thair souldiors, as said is, as they tender the saiftie and weill of the armie and success of the caus in hand.

(sic subscribitur,)

Burghly.
Cowpar.
Murray.
Capryngtone.
Craighall.
Edward Edgar.
Mr Wm. Moir.

Edict for the Committie of War at Kirkcudbrgt, for thair Accomptes.

At Edinburgh, the last day of November, 1640.—Forasmeikle as the Committie of Estaites of Parliament, with advyse of the Nobilitie and Commissioners for schyers and burrowes within this kingdome, have fund it expedient that the whole common burdens contracted for the use of the publict be knawn, and that the comptes of all Commissares, Collectores and uthers, intrometers with anie thing perteining to the publict, be cleirit and put in order. And for this effect, the Committie of Estaites have nominated certaine persones of the nobilitie and gentrie and burrowes to be auditors of the said whole accompts, and to mak ane formal report thairof, upon the xxij day of December next, conforme to ane act maid thairanent, of the date of thir presents. And sieing it is necessar that the haile common burdens, dews and accomptes foirsaid be prepared and in rediness for the view of the said auditors, betwixt and the tyme foirsaid. Quhairfore, the said Committie of Estaites of Parliament doe heirby warne, premoneis and requyer all Commissares and Collectores within the scheriffdome of Wigtone, and all uther persones within the said schyer, who have advancet or lent money, silver plate, victual, goodes, or anie uther commoditie quhatsumever, for the use of the armie or the publict, and who have been at anie trew and real charges in the effaires of this kingdome, and all uthers within the said schyer to whom the Estaites of this kingdome are adebted in anie soumes of money or uther particulares whatsumever, that they prepare thair comptes and present thame befoire the auditors appoynted be the Committie of Estaites, upon the first day of Januarij, with continuation of dayes, as the dyet appoyntit for heiring of the comptes of the scheriffdome foirsaid. With certificatione to anie persone to whom thair is anie thing adebtit to the publict, that if they failzie to give up anie thing restane to thame to the said auditores, appoynted be the said Committie of Estaites for that effect, they shall be halden as contumax and censurit thairfore be the said Committie, whensover they shall efterwards be desirous of satisfactione. And ordaines thir presents to be published at the mercate croce of the heid burghe and at ilk paroche kirke within the said schyer, that none pretend ignorance thairof.

(sic subscribitur,) Burghly.
Cowpar.
Murray.
Craighall.
J. Hamiltone.
Edward Edgar.
Mr Wm. Moir.

The Committie of the Stewartrie foirsaid, halden be a sufficient coram, at Kirkcudbryt, the first day of January, 1641. Alexander Gordon of Erlistone preses.

Act—Officeres of Kirkcudbryt.

The quhilk day, the Committie ordaines that the officeres of the towne of Kirkcudbryt have for thair attendance upon the Committie xx merks money. And to be peyit to thame be Robert Gordon of Knockbrax off the money that was peyit be David Makmollane as ane pairt of his fyne.

Act—Margaret Dumbar.

The whilk day, anent the supplicatione presented be Alexander Cavanes, in name of Margaret Dumbar, spouse to Gilbert Browne of Bagbie, desyering the said Committie to modefie and allow to her off her said husband's goodes and geir, for aliment of her and her childrene of ane competent meanes. Quhilk supplicatione, being heard and considered be the said Committie, they doe (dureing the space of twentie dayes to cum,) allow to her and her said childrene the number of thrie bolls meill, at viij pecks the boll. To be peyit to her be William Glendonyng, Commissar Depute, or anie uther whom he shall allow for that effect. With certificatione to the said uthers within the said schyer to whom the Estaites of this kingdome are adebted in anie soumes of money or uther particulares whatsumever, that they prepare thair comptes and present thame befoire the auditors appoynted be the Committie of Estaites, upon the first day of Januarij, with continuation of dayes, as the dyet appoyntit for heiring of the comptes of the scheriffdome foirsaid. With certificatione to anie persone to whom thair is anie thing adebtit to the publict, that if they failzie to give up anie thing restane to thame to the said auditores, appoynted be the said Committie of Estaites for that effect, they shall be halden as contumax and censurit thairfore be the said Committie, whensover they shall efterwards be desirous of satisfactione. And ordaines thir presents to be published at the mercate croce of the heid burghe and at ilk paroche kirke within the said schyer, that none pretend ignorance thairof.

(sic subscribitur,)

Burghly.
Cowpar.
Murray.
Craighall.
J. Hamiltone.
Edward Edgar.
Mr Wm. Moir.

Act in favores of Mary Murray and Bessy Geddas.

The quhilk day, anent the supplicatione presented be Mary Murray, spouse to Robert Maxwell of Culnachtrie, ante-covenanter, and Bessie Geddas, spouse to Harry Lyndsay of Rascarrell, also ante-covenanter, schawing that, whereas through the rebellione of thair said husbands, bothe against religione, covenant, king and countrie, the whole goodes, geir, cornes, cattle, rentes and uthers perteining to thame, and now to the publict, are sequestrate and appryset to the use of the public. Quhairby, thair said spouses, childrene and families are reducit to extreme miserie and hardness, not haveing quhairupon to sustein thame.—Desyering the said Committie to allot and allocate to thame and ilk ane of thame, viz.—to the said Mary Murray and Bessy Geddas, ane competent localitie, furthe of the redrest of thair said husbands' rentes, goodes and geir, for aliment of thame and thair said childrene, as the said supplicatione at length beires. The quhilk supplicatione being heard, sein and considerit, doeth heirby allot and allow to the said Mary Murray and Bessy Geddas, for aliment of thame and thair childrene, the victuall efter exprest, furthe of the redrest of thair said husbands' cornes and cropes, of the crop 1640, ilk ane of thame for awn pairts as is efter devydit, viz:—to the said Mary Murray, the number of ten laides of aites and ane boll beir, messor of Kirkcudbryt,—and to James Maxwell, father to the said Robert, the number of ten laides aites, messor foirsaid,—and to the said Bessy Geddas, the number of ten laides aites and ane boll beir. And that as ane competent allowance and mantenance to the persones respectively above named, for the crop and yeir of God foirsaid; and the said localitie to be frie of all publict burdene or uther burdene whatsumever.

Letter in favors of James Montgomerie.

Ryght Honourabill.—Sieing the Committie of Estaites hath appoynted me to lift the king's dewties and casualities within Galloway, I have maid bold to signifie the samen unto you, and to schow you that I have appoynted James Montgomerie, sone to the laird of Brigend, to uplift the samen. And for that effect I have appoynted him to address himselfe to your Lordship and your Committie, bothe above and benethe Crie, to concur with him to the poynding or anie uther way ye think fit. To the which doeing you do testifie your respect to the publict, and shall oblige your affectionate freind and servant,

J. C. Gaitgirth.

Postscript.—I hoipe ye will not onlie concur, bot be good example yourselffes.

Ayr, 10 Dec., 1640

For the Ryght Honourabill and my very good lord, Lord Kirkcudbryt, and to the Committie of War within his lordship's divisione.

Act—Gaitgirth.

The quhilk day, anent the letter presented be James Montgomerie frae the laird of Gaitgirth, schawing that the said laird hes right frae the Committie of Estaites for intromissione with the king's rentes within Galloway, and that he hes constituted the said James Montgomerie uplifter thairof. The said Committie, takeing the samen to thair consideratione, and finding that the Commissar Depute within thair bounds hes upliftit ane great pairt of the said rentes else, and that it wald mak ane confussion in his accomptes if he should desist thairfrae; thairfore ordaines the said Commissar Depute to proceid and uplift the said rentes and be comptable thairfore to the said laird of Gaitgirth.

Act—Commissar Depute.

The quhilk day, the Committie requyerit James Montgomerie to delyver to the Commissar Depute ane rentall of the king's rentes within his divisione. The said James altogither refuissit to doe the samen. Quhairupon the said Commissar Depute requyerit act.

Act for Accomptes.

The quhilk day, anent the drawing up of the accomptes of what is debursit in the publict effaires in this divisione, and to be presented to the Lords, auditores appoynted for the said accomptes; the Committie ordaines Alexander Gordon of Erlistone, Johne

Fullartone of Carletone, Richard Mure of Cassincarie, Robert Gordon of Knockbrax, and George Glendonyng of Mochrum.

Act contra Willsone.

The quhilk day, Johne Willsone, in Cumstone, becumes cautioner for William Murray, souldior, that he shall goe to the armie and purchaise his Captaine's note of his receipt; and that under the paine of ane hundred merks, to be peyit be him to the Collector of the said Committie.

Act—James Maxwell of Brekansyde.

The quhilk day, the said Committie, sieing that James Maxwell of Breckansyde his whole rent comes in to the use of the publict, except his wyfe's competencie; ordaines that the Commissar Depute within that divisione, off the first end of the rent and dewtie of the said James his landes of Barnecrosche, deduce the tenth and twentieth penny rent and trowp horss money, for the said James his whole landes on this syde the watter of Urr, provydeing the said tenth and twentieth penny rent and trowp horss money be peyit.

The Committie of the Stewartrie foirsaid, halden be a sufficient coram, at Kirkcudbryt, the secund of January, 1641. Erlistone preses.

Act—James Maxwell of Brekansyde.

The quhilk day, sieing that James Maxwell of Brekansyde, ante-covenanter, hes fund suretie to William Glendonyng, Commissar Depute, for peyment to him, betwixt and the last of this moneth, of the pryce of his whole moveable goodes and geir, conforme to the band maid thairanent, of the date of this present act; and that the said James is dew furthe of the landes of Buittle, to the publict, certaine teindes and tak dewties quhilk perteinit to the pretendit Bischope of Edinburgh, for certaine yeires bygane; and also is dew to the minister certaine bygane teindes of the said landes; quhilk the said James cannot be lyable to pey, sieing his haile rent and moveables is peyable to the publict. Thairfore ordaines the said William Glendonyng, at the ressait of the soume conteinit in the said band, to grant ane discharge to the said James of the teind tak dewtie, the crop and yeir of God foirsaid; and also the bygane teindes dew to the minister the said yeires; and keip the samen frie thairof. And the said William to compt with the publict for the said teind tak dewtie, the said yeires, and for the superplus of the money conteinit in the band, the minister's dew also being peyit.

Act—Richard Mure of Cassincarrie.

The quhilk day, the Committie taking to consideratione that the dryveing of geir to be appryset for peyment of what is dew to the publict, that cannot presentlie be gotten sauld, is much the worse. Thairfore, ordaines that it shall be sufficient to the Captaine of the paroche, in presence of four honest men, to appryse the samen allanerlie upon the ground of the land.

And ordaines Richard Mure of Cassincarrie to poynd Johne Browne, merchand, his scheip, for what he is dew to the publict, at xxxs. the peice, and to keip thame upon the ground of the land, to the use of the publict; or else to sell thame on the pryce foirsaid, and satisfie the publict for what is dew.

Act—Roger Maknacht of Kilquhennattie.

The quhilk day, anent the supplicatione presented be Roger Maknacht of Kilquhennatie, schawing that, whereas, he is charged be the Commissar Depute for peyment of xij libs yeirlie more nor he is dew for the teindes of his landes, and quhilk is more nor the rental condiscendit upon be Johne Maknacht of Edinburgh, his uncle, and the pretendit Bischope of Edinburgh for his landes; quhilk he can prove if he had ane tyme for that effect, as the said supplicatione beires. Ordaines the Commissar Depute to ressave the rest of his said teindes, and to continew the peyment of the said xij libs yeirlie, the yeires restane, unto the next terme, whill Kilquhennatie proves the rentall.

Act—Maister Johne Makclellane.

The quhilk day, anent the supplicatione presented be Mr Johne Makclellane, Minister at Kirkcudbryt, schawing that, whereas, thair is dew to him furthe of the vicarege and gleib of Dunrod and out of the Mylnetone the soume of ane hundred punds money, as the said supplicatione beires. The quhilk being heard, sein and considered, decernes and ordaines Agnes Gordon, spouse to Robert Makclellane of Nuntone, ante-covenanter, to pey to the said Mr Johne, for the halfe of the said gleib and vicarege teindes, the soume of fiftie punds money, this crop, 1640. And ordaines the said Mr Johne to have furthe of the said landes of Mylnetone aught bolls victuall, halfe meill, halfe beir, and that for peyment of the uther halfe of the said hundred punds for the said gleib and vicarege teindes, and for the parsonege teindes of the said lands of Mylnetone.

Act—Penrie contra Macmollane.

The quhilk day, anent the supplicatione presented be Johne Gordon of Barskeoche, in name and behalf of Johne Penrie, indweller in the burghe of Ayr, schawing that, whereas, John Makmollane of Brocklock, be his band and obligatione subscryvit in his hand, of the date, at Dalmellingtone, the xiij day of Junij last bypast, for the causes thairin specifeit, band and obleist him to have contentit and paid to the said Johne Penrie, his aires and assignees, the soume of fiftie punds money of this realme, betwixt the date foirsaid of the said band and the first day of November thairefter next to cum, now bypast, with the soume of sextene punds money foirsaid of liquidate expenses. In caice of failzie, with the ordinar annual rent thairof, conforme to the act of Parliament, aye and whill the peyment of the said principal soume, efter the said terme, as the said supplicatione beires. Notwithstanding whereof, the said Johne Makmollane will on naewayes mak peyment to the said Johne Penrie of the soumes foirsaid, conteinit in the said band, but altogither postpones and defferes to doe the samen without he be compelled. Desyering, thairfore, the said Committie to give thair power and warrand to compell the said Johne Makmollane, in manner and to the effect above conteinit, as the said supplicatione togither with the said band in thameselffes more fullie proports.

The quhilk supplicatione and band being heard, sein and considerit be the said Committie, they decerne and ordaine the said Johne Makmollane to content and pey to the said Johne Penrie the soumes of money above conteinit, principall and expenses conteinit in the said band, efter the forme and tenor thairof, in all poyntes. And that Alexander Gordon of Knockgray, Captaine of the paroche quhair the said Johne M'Mollane dwelles, by himselfe or his constables, mak intimatione to the said Johne Makmollane to content and pey to the said Johne Penrie the soumes of money above specifeit, conteinit in the said band, efter the forme and tenor thairof and thir presents in all poyntes, within sex dayes next efter the intimatione. The quhilk sex dayes being bypast and nae peyment maid, that the said Alexander and his foirsaids poynd and distrenzie the said Johne Makmollane his redrest moveable goodes and geir whatsumever, quhairever the samen can be apprehendit within this Stewartrie, and mak the said Johne Penrie to be compleitlie peyit of the said soumes, principall and expenses, conteinit in the said band, efter the forme and tenor thairof and act foirsaid, in all poyntes. Becaus the said band was producit in presence of the said Committie, beirane the forme and tenor above specifeit.

Act be the Committie of Estaites for a mantenance for Brekansyde's Wyfe.

At Edinburghe, the first day of November, 1640 yeires Anent the supplicatione given in to the Lords and uthers of the Committie of Estaites of Parliament be Margaret Vans, spouse to James Maxwell of Brekansyde, making mentione, that in respect the said James Maxwell, her husband, being ane noncovenanter, she is informed that the said Committie of Estaites hes given or intends to give warrand for intrometting with his rentes for the use of the publict, whereby the said supplicant and hir four childrene will be destitute of mantenance, unless the said Committie of Estaites allow and dispense with ane pairt of her said husband's estaite for the mantenance of hir and hir childrene. Desyering thairfore, ane warrand and licence for intrometting with and uplifting frae the tenants of hir said husband's landes, of ane competent portione of hir said husband's estaite, for the mantenance of hir and hir childrene, conforme to the common order used and granted in favores of uthers in the lyke caice and conditione; as the said supplicatione at more lengthe proports. Quhilk supplicatione, togither with ane rentall of the said James Maxwell, her husband's rent and estaite, togither with ane note of hir debtes, given in be the Vyscount of Kenmore and subscryvit be his lordship's hand, being heard, read, advyset and considdered be the said Committie, the lords and uthers of the said Committie of Estaites doeth heirby modefie and allow to the said supplicant, for the aliment of hirselfe and hir said childrene, the soume of aught hundred merks money of this realme yeirlie, to be peyit furthe of hir said husband's rent and estaite, to be peyit to hir aither be sik tenants as shall be alloted and named be the Committie of War of Galloway, or utherwayes be the Commissar Depute of the bounds quhair the said landes lyes, furthe of the redrest of hir said husband's estaite.

Lykeas, these gives power to the said Committie of War to settle the foirsaid localitie, quhairby the said supplicant may be peyit thairof. And ordaines the remanent of hir said husband's rentes and estaite to be intrometted with be the Commissar of that schyer, and to be maid furthe cummane for the use of the publict, and for peyment of annuells to just creditores, who are covenanters, conforme to the warrand and order to be given be the said Committie of Estaites to that effect.

(sic subscribitur,) Burghly.
Cowpar.
Capryngtone.
Gaitgirth.
Robert Moir.
Richard Maxwell.
Mr Wm. Moir.
James Scott.

Act—Competencie to Margaret Vans.

The whilk day, the Committie of the Stewartrie of Kirkcudbryt, anent the warrand presented to thame be Margaret Vans, spous to James Maxwell of Brecansyde, ante-covenanter, frae the Committie of Estaites, of the date the first day of November last bypast, for allotting and allowing to the said Margaret, for hir and hir childrene thair aliment and mantenance, of the soume of aught hundred merks money of this realme yeirlie, to be peyit furthe of the redrest of hir said husband's rentes and estaites, as the foirsaid warrand of the date foirsaid at length beires. In obedience to the said warrand, and efter tryall taken of the said Margaret's husband's estaite and rent underwrytten and uthers, doeth heirby allot and allow to the said Margaret, for aliment of hir and hir said childrene, in peyment of the foirsaid soume of aught hundred merks money yeirlie, the landes and uthers efter specifeit, mailles, fermes and dewties thairof, viz.—all and haill the landes of Meikle and Little Brekansyde, the Halfe-merk-land of Lochend, the landes of Whytesyde, the Fourtie-schilling-land of Porterbellie, howses, biggings and haill pertinents of the said landes, all lyane within the parochen of Kirkgunzeon and Stewartrie of Kirkcudbryt; and alsua, all and haill the landes of Little Slachtes, lyane within the parochen of Kirkmabreck and Stewartrie foirsaid; and lykewise, furthe of the redrest of the mailles, fermes and dewties of the Seven-merk-land of Barnecrosche, lyane within the parochen of Tongland and Stewartrie foirsaid, the number of seven bolls beir and six bolls meill, at viij peck the boll, messor of Kirkcudbryt; quhilk will fullie compleit the foirsaid localitie of aught hundred merks money yeirlie.—With power to the said Margaret to intromet with and uplift the foirsaid localitie, allotted as said is, in manner and for the causes above specifeit, and that yeirlie, as is above conteinit, and, if neid beis, to call and pursew thairfore, as accords of the law.

Act contra Newall.

The quhilk day, the Committie taking to thair consideratione that non-covenanters cannot be lyable in peyment of teindes, few dewties, or other dewties to the publict, and that thair moveable goodes be applyet to the use of the publict. Thairfore the Committie discharges Johne Maxwell of Newlaw for charging or molesting of the persones underwrytten for peyment to him of what they are dew to the publict, and that sicing thair whole rents and moveables is to be intrometted with be the Commissar Depute. They are to say—Robert Maxwell of Cullnachtrie, Harry Lindsay of Rascarrell, William M'Clellane of Overlaw, William M'Clellane of Ards, David Cairnes of Kip, Johne Tait of Castell, and Gilbert Browne of Bagbie.

1 The Cairnes were proprietors of various lands in the Stewartry at an early date, and the lands of Kipp only passed out of their hands about sixty years ago, when they were sold by the late George Cairns, Esq. of Kipp, a most eccentric and well known character in Galloway, who died about 1804-5.

The following curious extract from an agreement between Thomas M'Clellan of Bombie and Edward Cairnes of Auchengule, called Barron Cairnes, dated the last day of October, 1588, is taken from the Register of Deeds, and will give a good idea of the manner which lands were then taken.

"For the quhilk set of the lands foirsaid, [Netherthirde,] the said Thomas M'Clellan of Bombie, as principall, William M'Clellan of Balmangane, Matho M'Ilwraith in Little Stockarton, Henrie Cotrane in Black Stockarton, and Johne Thomsoun in Gribdie, as cautioners and suirtie for said Thomas, bind and obleis thame, thair aires and assignayes, conjunctlie and severallie, to content and pay to the said Edward yeirlie, during his lyfetyme, the soume of four score punds money yeirlie at twa termes in the yeir, Witsounday and Mertinmes, be equall proportiones, the first

Act—Ewart.

The quhilk day, anent the supplicatione presented be Patrick Makclellane, notar, and Robert Ewart, burges of Kirkcudbryt, shawing that, quhairas, Johne Ewart, baillie of the said burghe, upon the xx day of October last, cam to the said Patrick his chalmer and borrowit frae him the legall testament of the umq^le. Johne Ewart, his good-father, and faithfullie promeissit to have redelyverit the samen bak again uncancellit within ane hour thairefter, quhilk he on nae wayes will do as yet; and thairfore desyering ane warrand to be given be the said Committie to delyver the said testament, that the samen may be confermit, as the said supplicatione beires. The quhilk being heard, sein and considderit be the said Committie, and the said parties thair rights and alledgances being heard and compearane, ordaines the said testament to be registerit be the Clerk of the Committie in the books thairof, word be word, and the principall delyverit be the said Clerk to Robert Forrester, Commissar, to be comfirmed be him as Commissar; quhairof the tenor follows.

termes payment being at Mertinmes, in the yeir of God J^m V^c four score nine yeires, and fyve bolls beir and fyve bolls of meill, small messor, yeirlie, betwixt Yule and Candilmes, the first yeires payment being betwixt Yule and Candilmes, in the said four score nine yeires. * * * * And also, the said Thomas bindes and obleiss him and his foirsaids to gif yeirlie to the said Edward, during his lyfetyme, ane haill new garment and cleithing to his haill bodie, of sic claith as the lady Bombie, his wyfe, sall yeirlie maik, and sall manteine and defend the said Edward as his speciall serwand, in all his affaires as he sall haif ado."

1 See Appendix.

At Kirkcudbryt, the xiij day of Maij, Jm. VIc. and fourtie yeires, befoire thir witnesses, Johne Ewart younger, burges of Kirkcudbryt, Andro and Johne Ewartes, his sones, Johne Ewart in Creikheid, Adam Ewart, burges thair, Thomas Ewart, burges thair, Johne Makjore, Johne M'Michein in Grange.

Item,—He nominates, constitutes and appoyntes Robert Ewart, his sone, and Helene Ewart, his dochter, his onlie executores and intromettores with his haile goodes and geir; and gives and commits to thame full power to give up all debtes bothe in-awing and out-awing to him and be him to uthers. This done day, yeir and place foirsaid, in presence of the witnesses abovewrytten.

(Sic subscribitur,) I, Johne Ewart, with my hand at the pen led be the notar underwrytten, at my command, becaus I cannot wryte myselfe. Ita est, Patricius Makclellan, notarus publicus, de mandato dicti Johanis scribere nesciens ut asertit, manu mea.

APPENDIX.

APPENDIX.

M'Kie of Larg.

The M'Kies of Larg were descended from one of three brothers, M'Kie Murdoch and M'Lurg, to whom the thirty pound land of the Hassock and Comloddan was granted by Robert Bruce, as a reward for the sevices which they had rendered him during his struggles with the English for the independence of Scotland.

The following account of the manner in which they acquired the land was written by Andrew Heron of Burgally, and was first published in the Appendix of Symson's Description of Galloway.

"King Robert, being by a part of the English army defeat in Carick, fled into the head of Loch-die to a few of his broken partie, and lodging in a widow's house, in Craigencallie, in the morning she, observing some of his princely ornaments, suspected him to be a person of eminence, and modestly asked him in the morning, if he was her Leidge Lord. He told her Yes, and was come to pay her a visit; and asked her if she had any sons to serve him in his distress. Her answer was, that she had three sons to three severall husbands; and that if she was confirmed in the truth of his being their sovereign, they should be at his service. He askt her farther, if she could give him any thing to eat. Her answer was, there was litle in the house, but agust-meal and goats'-milk, which shou'd be prepared for him; and while it was making ready, her three sons did appear, all lusty men. The King askt them, if they wou'd chearfully engage in his service, which they willingly assented to; and when the King had done eating, he askt them what weapons they had, and if they could use them; they told him they were used to none but bow and arrow. So, as the King went out to see what was become of his followers, all being beat from him but 300 men, who had lodged that night in a neighbouring glen, he askt them if they could make use of their bows.—M'Kie, the eldest son, let fly an arrow at two ravens, parching upon the pinnacle of a rock above the house, and shot them through both their heads. At which the King smiled, saying, I would not wish he aimed at him. Murdoch, the second son let fly at one upon the wing, and shot him through the body; but M'Lurg, the third son, had not so good success.

"In the meantime, the English, upon the pursuit of K. Robert, were incamped in Moss Raploch, a great flow on the other side of Die. The King observing them, makes the young men understand that his forces were much inferior. Upon which they advised the King to a stratagem, that they would gather all the horses, wild and tame, in the neighbourhood, with all the goats that cou'd be found, and let them be surrounded and keept all in a body by his soldiers in the afternoon of the day, which accordingly was done. The neighing of the horses, with the horns of the goats, made the English, at so great a distance, apprehend them to be a great army, so durst not venture out of their camp that night; and by the break of day, the

King with his small army, attacked them with such fury, that they fled precipitantly, a great number being killed; and ther is a very big stone in the centre of the flow, which is called the King's Stone to this day, to which he leaned his back till his men gather'd up the spoil; and within these thirty yeares, there were broken swords and heads of picks got in the flow, as they were digging out peats.

"The three young men followed close to him in all his wars to the English, in which he was successfull, that at last they were all turn'd out of the kingdom, and marches established 'twixt the two nations; and the soldiers and officers that followed him were put in possession of what lands were in the English hands, according to their merite. The three brothers, who had stuck close to the King's interest, and followed him through all dangers, being askt by the King, what reward they expected? answered very modestly, That they never had a prospect of great things; but if his Majesty would bestow upon them the thirty pound land of the Hassock and Comloddan, they would be very thankfull; to which the King chearfully assented, and they kept it long in possession."[1]

The lands thus acquired were divided amongst the three brothers,—M'Kie obtained the Larg, Murdoch, the Risk, and M'Lurg, Machermore.

---

1 "There are no lands called *Has ock* in the grant made by the King. The oral tradition of the country is, that Annabel, the widow, solicited and received, the bit *hassock* of land that lies between the burn of Palnure and the burn of Penkill." This *hassock* of land is an isosceles triangle, the base of which runs for three miles along the Cree, and the sides formed by the streams of Palnure and Penkill, run five miles into the country. This speck of land has been the birth place or residence of more distinguished individuals than, perhaps, any other rural spot of equal extent in Scotland. Macmillan, the founder of the sect that bear his name, was born at Barn-

Y*

When the army of the Covenanters took the town of Newcastle from the King's forces, in 1640, the troops commanded by Sir Patrick M'Ghie or M'Kie of Largo in Galloway, were particularly distinguished. In the engagement, however, Sir Patrick lost his only son, a brave aspiring youth, who was standard bearer to Colonel Leslie's troop. He was the only person of any note who fell on the side of the Covenanters, and was much lamented by the whole party. Zachary Boyd, in a long poem, entitled "Newburn Book, thus deplores his death;—

> "In this conflict, which was a great pitie,
> We lost the son of Sir Patrick M'Ghie."

The Stewartry of Kirkcudbright has been represented in Parliament by various members of this family, viz;—Alexander M'Kie of Palgown, who was a member of the Scottish Parliament at the time of the Union; John M'Kie, who sat in the British Parliament from 1747 to 1754; and J. Ross M'Kie, who represented the said county from 1754 to 1761. John M'Kie, Esq., of Bargally, the present member of Parliament for the Stewartry of Kirkcudbright, is also descended from the M'Kies of Palgown.

cachla. Murdoch, the last of the descendants of old Annabel, who was settled in Kirouchtree, was famed over Europe for his knowledge of Botany. Dr William M'Gill, minister of Ayr, whose Essay on the death of Christ caused so much controversy near the close of last century, received the greatest part of his education at the school of Minnigaff.—Alexander Murray, late Professor of Oriental Languages in the College of Edinburgh, was born at Corwar; Patrick Heron, whose Banking scheme ruined many gentlemen in Galloway and Ayrshire, occupied Kirouchtree; and Lieutenant General Sir William Stewart, who fought so bravely under the Duke of Wellington, possessed the estate of Comloddan, all within the King's grant to Annabel."—HISTORY OF GALLOWAY.

Carsphairn.—Mr Johne Semple.

Carsphairn was only erected into a parish a few years before 1640; previous to this time that part of Carsphairn which lies on the east side of the water of Deuch formed part of the parish of Dalry, and that on the west side part of Kells; both of which parishes still receive a stipend of about £9 sterling from Carsphairn.[1]

By a charter granted, to Robert Grierson of Lag, in 1671 and ratified in parliament in 1672, the village near the church of Carsphairn was created a free burgh of the barony to be called the Kirktoun, with power to elect bailies and other officers, to build a tolbooth, and a cross, to create burgesses, and to hold a weekly market and two annual fairs.

1 "Dec. 16th, 1638. Then there was a Supplication presentit in name of the Kirk of Corspairne, which church lyes in a very desolat wildernes, containing 500 communicants. It was builded by some gentlemen to their great expenses, only out of love to the salvation of the soules of a number of barbarous ignorant people who heirtofoir hes lived without the knowledge of God, their children unbaptized, their died unburied, and could find no way for getting mentainance to a minister but to the sympathizing of zealousness as the Assembly would think expedient.

"My Lord Cassiles said—Their cace is verie considerable, and deserves helpe. The cace of their soules is verie dangerous, being 15 or 16 myles from a church; and now, since God hes given them the benefite of a kirk, I think verilie a very little helpe of the Presbitries of the kingdom would give them a competent meanes for a minister, especiallie seeing they have alreadie provydit something themselves.

"Dec. 17th. After in calling upon the name of God, those who were appoynted to meit about the Kirk of Carsfairne, declaired that they had mett and taken consideration of the estate of the kirk; and, finding that the pairties that possesses the teynds cannot be moved to give provision; and, we thinke it expedient they be helped ane uther way; and becaus we thinke it expedient that the whole kingdome be not troubled with it; therefore we thinke the bounds of this syde of Tay, including Fyfe and Forthe, will be sufficient."—Records of the Kirk of Scotland.

Mr John Semple who was long minister of Carsphairn, was allowed by the covenanters to be possessed of the gift of prophecy. He died at Carsphairn about 1677, and was suceeded by Mr Gilchrist. From various entries in the Burgh Records of Kirkcudbright, it appears that he left a considerable sum of money to the poor of that burgh.

Shortly after the Revolution, a curious pamphlet entitled, "Some Remarkable Passages of the Life and Death of Mr John Semple," was compiled by Patrick Walker, the packman, the writer of Peden's Life and Prophecies. As this work is now very rare we have subjoined a few short extracts from it.

"Mr Semple by his singular Piety, and examplery Walk, was had in such Veneration, that all Ranks, and Sorts of People, stood more in awe of him than many Ministers, yea, he was a great Check upon the lazy corrupt Part of the Clergy, who were much afraid of him. He was very painful and laborious amongst his own People, preaching frequently on Week days, which is now rarely done in Country Places. The Lord's Presence with him in Preaching, Catechising, and in the Exercise of Church-disipline, reclaimed the People, who were scarcely civilized before; Several of whom became eminant Christians, and were endued with the Grace of Prayer, of whom Mr Peden used to say, That they had Moyen at the Court of Heaven, beyond many Christian Profesors of Religion he knew.

"He used to wait very carefully upon Church-judicatories, and very rarely was absent, and that from a principle of Conscience, tho' Carsphern be twenty four miles distant from Kirkcudbright the Presbyterie's Seat, notwithstanding that much of the Way is very bad. When he was going to the foord in the Water of Dee, in his way to the Presbytery, he would not be hindred from riding the Water, tho' he was told by some, that the Water was unpassible, saying I must get through, if the Lord will I am going about his work, He entered in, and the Strength of Water carried him and his Horse beneath the Foord, he fell from his Horse, and stood up in the Water, and taking off his Hat prayed a Word to this Purpose, Lord, art thou in earnest to drown me thy poor Servant, who would fain go thy Errands? After which, he and his Horse got both safely out, to the Admiration of all Onlookers.

" When a neighbouring Minister was distributing Tokens before the Sacrament, Mr Semple, standing by and seeing the Minister reaching a Token to a Woman, said Hold your Hand, that Woman hath got too many already, for she is a Witch, of which none suspected her then, yet afterwards, she confessed herself to be a Witch, and was put to Death for the same.[1]

"A little before his Death, he was apprehended, and after nine Months Imprisonment in the Castle of Edinburgh, was taken before the Council for his Nonconformity, they threatned him severely with Death or Banishment, he answered with Boldness, He is above that guides the Gully,

1 This woman most probably was Elspeth M'Ewen, who was burned for witchcraft at Kirkcudbright in 1698. Between the years 1630 and 1700 there were several other persons in the Stewartry of Kirkcudbright tried for witchcraft, of whom, some were sentenced to be whipped, and others to be banished from the bounds of the Stewartry; but Elspeth appears to have been the only one who was executed. At this time the reality of witchcraft and sorcery was never called in question, and the number of trials and executions which took place in Scotland are almost incredible. Balfour in his Annals of Scotland says, " Many witches were apprehendit and commissions being giuen by Parliament and the counsell for their tryell, they were execut in the shyres of Fyffe, Perth, Stirling, Linlithgow, Edinburgh, Haddintone, Mersse and Peibles, &c. I Mayselue did see, on the

my God will not let you either kill me or banish me, but I will go home and die in Peace, and my Dust ly among the Dust of the Bodies of my People; accordingly the Council dissmissed him.

"After this he went home, and entered his Pulpit, he said I parted o'er easily with thee, which has been many a sore Heart to me, but I shall hing by the Wicks of thee now; and on his Death bed, his Zeal and Concernedness, for the Salvation of his People was such, that he sent for them, and preached to them, freely shewing them what Danger their Souls were in by Reason of their Unbelief and Estrangement from the Power of Godliness; laying before them their manifold Sins; to make them sensible of their Need of Christ, expressing great Fears, that he would give up his Accounts, as to many of them with Grief. Which Words were so accompanied with Power, that made many of them weep bitterly, which would be a wonder in this hardened and obdurate Age, and mocked at, as the only Effects of a silly waterish Constitution, as if all the Tempers of the saints were flashy, such as David, and Peter, yea, and Christ himself, in whose Constitution there was nothing defective, who did frequently weep.— However Mr Semple's Weepers were not all of the flashy

20 of Julij, this zeire, [1649,] in one afternoone, commissions seuerallie directed by the parliament, for traying and burning of 27 witches, women, and three men and boyes; ther depositions wer publickly read in face of parliament, before the house would wotte to the president's subscriuing of the acte for the clerke issewing of these commissions; Lykwayes diuers commissions wer giuen by the Lords of Counsell, in Nouember and December, this same zeire, for traying and burning of witches; ther depositions wer read, amongst the wich ther was one that confessed that she had bein of lait at a meitting with the deuill, at which ther wer aboue 500 witches present. So far had that wicked enimy of mankind prewailled, by his illusions and practises, one these poore wretched miserable soules."

kind, for many af them proved solid Christians, and lived to acquit themselves Men and Christians on proper Occassions.

"He died with much Assurance of Heaven, and longed to be there, rejoicing in the God of his Salvation; and under great Impressions of dreadful Judgements to come on these covenanted Lands, especially on Scotland, and the West and South thereof, above all other places, by the bloody Sword of Popish and others taking Part with them; repeating these Words three Times over, A BLOODY SWORD FOR SCOTLAND. He was buried in the Churchyard of Carsphern, and it is said, his grave is known there to this Day."

Gordon, Viscount Kenmure.

Whence the origin of the Gordons, who were one of the most ancient and powerful families in Scotland, there are now no means of accurately ascertaining; some historians, reasoning from the similarity of names, have alledged that the first of the name came from Gordonia, a city in Macedon, whilst others trace them to Normandy, where there once was a manor called Gordon, and conclude them to be sprung from the same family as Bertrand de Gordoun, the archer who shot Richard I. at the siege of Chalos in Acquitaine. The traditionary account of the origin of the name is that, in the reign of Malcolm III, there was in the South of Scotland a wild boar of tremendous strength and ferocity which had killed many knights and gentlemen who had attempted to destroy it, and had at length become such a terror to the whole country that none dare to encounter it, whereupon the king offered a great reward to whoever should kill it and bring its head to the Court. This being done by a brave yeoman, called Adam, the king enquired at him how he slew the monster, he replied, that having wrapped his plaid about his arm he thurst it into the mouth of the boar and *gored him doun* with his dagger. Malcolm, pleased with the intrepidiy of the action and ingenuity of the device, conferred upon him the honour of knighthood and commanded him to assume the surname of *Goredoun* in commemoration of the circumstance. By some the boar is said to have been killed in the Forest of Glenkens, whilst others lay the scence of the exploit in the parish of Gordon in Berwickshire.

Among the first members of this family who make any considerable figure in Scottish history is Adam de Gordon, who was one of the chief commanders of the auxiliary forces sent by Alexander II. to accompany Louis, king of France, in his expedition to the Holy Land in 1270. His grandson, Sir Adam de Gordon, who was one of the greatest men of his time and a friend and companion of Sir William Wallace, having assisted that patriot in his expedition into Galloway, in 1297, was by him made Keeper of the Castle of Wigtown, which they had wrested from the English. He was slain at the battle of Halidon Hill in 1333. From his eldest son, Alexander, are descended the Gordons in the North, afterwards raised to the Peerage with the title of Dukes of Gordon; and from his second son, William de Gordon, is sprung the family of Stitchell and Lochinvar.

The Gordons of Lochinvar, one of whom fell at Flodden and another at Pinkie, are often mentioned in history as taking a leading part in the Border Wars.

Sir John Gordon of Lochinvar, a great loyalist and steady adherent of Charles I., was in 1633 created a peer by the title of Viscount Kenmure, Lord Lochinvar, &c. He also obtained a charter erecting part of his lands into a royal burgh, with ample jurisdiction, to be called the burgh of Galloway, now New-Galloway. He died at Kenmure in 1634, and a circumstantial account of his pious demeanour upon his death-bed, is given in a small work, supposed to be written by Rutherford and entitled, "The Last and Heavenly Speeches and Glorious Departure of John Viscount Kenmure." In 1628 he married Lady Jean Campbell, daughter of Archibald, 7th Earl of Argyle, and by her had a son,

John, 2nd Viscount, who died under age, in 1639, and was succeeded by his cousin,

John Gordon of Barncrosh, 3rd Viscount, who dying in 1643, was succeeded by his brother,

Robert, 4th Viscount. This nobleman suffered great hardships on account of his loyalty and was one of those who were excepted out of Cromwell's act of grace and pardon; his estates were forfeited, and his Castle of Kenmure having been taken by a party of Cromwell's troops, was burned. He survived the Restoration and died at Greenlaw in 1661, when the title devolved upon

John Gordon of Pennygame, 5th Viscount, who dying, in 1663,[1] was succeeded by his brother,

---

1 As it may be interesting to know in what manner a nobleman's house was furnished at this time, we have subjoined an Inventory of the Furniture, &c., which was in the house of Greenlaw at this nobleman's death.

At Greenlaw, the penult day of Februarij, J^{m} VI^{c} sextie thrie yeires, (1663,) the insight of the place of Greenlaw is inventired beffoir the persones following, they are to say—Robert Mr. of Herreis, Alexander Gordoun of Penninghame, William Gordoun of Shirmeres, Roger Gordoun, appeirand of Troquhaine, Johne Maxwell of Breckensyde, Alexander Maxwell his brother, and Robt. Gordoun of Grainge; which inventar underwrytten is wrytten be Johne Hutoune, younger, notar.

Imprimis, in the kitchen chamber, ane stand bed and ane draw bed, both

Alexander, 6th Viscount, who had command of a regiment at the battle of Killicrankie, where the most of his men were killed. He died in 1698, and was succeeded by his son,

corded; a suit purple velvet courtinges, conteining fyve peice, with a pan conforme thereto, all with gold lace, lined with yellow taffitie, with the inner valenes and heid of yellow taffitie; ane table cloth conforme to the courtinges, with a gold fringe going about it; with ane arm chair, two stooles and ane foot gange conforme to the bed; Lykewayes, within the stand bed, ane cours fethir bed, twa pair of cours blanketts, ane code, a couering thereof, a pair of cours sheites, and ane auld mat wich was taffitie; and in the draw bed, ane pair of cours sheites and blankets and ane blew whyt and yellow mat; with twa timber tables. Item, in the parlour chamber, ane large stand bed and draw bed, both corded; a suit of sad serge courtinges, conteining four peice, with pan and coverlett conforme thereto, the pan haveing a deep orange silk fringe, the taster and inner valenes haveing ane orange fringe and heid peice all of the lyning of the courtinges; thair is within the bed, a caffe bed, fethir bed, bolster, twa cods and couerings thairof, pair of sheites, twa pair of blankets, twa armed chaires, four stooles, conforme to the bed; with a green table cloth with a lytle imbroderie about and a green fringe; with twa timber tables; the draw bed thair are nothing in it, with dry stool and pot. Item, in the chamber above the parlour, ane stand bed, corded, with thrie peice of red taffitie courtinges, with pan, a silk fringe conforme; thair is in the bed, fethir bed, bolster, cod and covering thairof, and ane pair of blankets; ane table cloth of strype hingings; with ane chair conforme to the table cloth, and ane pot. Item, in the chamber next adjacent to the chamber above the parlour, ane stand bed with a draw bed, both corded, with courtinges, coverlet and pan, all of gray serge, thair being four peice of the courtiniges and the pan haveing a gray silk fringe; thair is in the bed, a caffe bed, a fethir bed, a pair blankets, and a red worset rug; with thrie chaires covered with strype hingings; ane auld grein table cloth with yellow border about it. Item, in the lytle chamber, a stand bed, corded; a fethir bed and bolster; a pair of blankets; a red worset rug; thrie peice of auld taffitie hingings; a pan of blew cloth and broidered with black velvet, with a blew and whyt silk fringe, wanting a taster. Item, in the chamber next to the lytle chamber, a chapell bed all of bundwork with two peice of hingings and pan, all of flowrit din danes of silk, and a roustie silver fringe; thair is in the bed, a kynde of fethir bed and bolster, a pair of blankets and roune sheites, and a countrie covering; ane chair covered with strype hingings. Item, in the dyning roume, a reasonable long oak table with a large carpet upon it; another owile table of sweet wood and a carpet upon it, conforme with ane uther lytle tymber table and sevintein chaires covered with red

William, 7th Viscount, who raised the standard of the Pretender at Lochmaben, in 1715, and had the chief command of the rebel forces in the South. Proceeding with

Rushew leathir and brass nails; a pair of table chakeres and men conforme; with a going knock and knockcaice; with tonges and fyre shuffell; with ane lytle copbord within the wall. Item, in the kitchen, fourtein hounsell plaits; ane pair of great lyning raxs; four spits; with ane great pot, meidle pot, and ane lytle pot; twa panes; a morter; with ane frying pan; a droping pan; a brandereth of iorn; ane ladle and fork. Item, in the servant men's house, a kist; with ane auld copboord; four brass common candle sticks, twa tin skakats; twa quart stoypes; with a large silver salt fat; ten silver spoones; a caice and ten kniffes. Item, in the lairner, ane mat and ane pair of blankets; ane meikle kist; seven hogheids and twa barrelles. Item, in the woman house, twa peice of red hingings with blew and whyt lace and pan of the samyn going about the half of the stand bed, and twa pair of auld blankets; and ane uther turnit bed, with thrie peice of flowerit courtinges and pan conforme to it, with ane couerlet of strype hingings; ane auld table; a lame pot for watering chamberes. Item, in he work-house chamber, two beds; a pair of blankets; a pair of sheites; with a table and two formes. Item, in Margaret Ramsay's bed chamber, two pair auld blankets and two beds. Item, in the garnell house, twelff great Inglisch pewder plaites, thrie petingires, twelff trencheres and four salsters, all haveing upon them my lord's armes; four Inglisch candle-sticks of Inglisch pewder; twa Inglisch flagones, lykwayes haveing upon them my lord's armes; with ane chamberpot and twa Scotes candle-sticks. Item, in the chamber above the kitchen chamber, a large stand bed and a draw bed, both corded, with compleit hingings about the bed all of one peice, with pan and taister and table cloth, all of strype hingings; thair is in the bed, a fethir bed, a bolster, twa pillowes and pillow beirares, with a pair of ordinarie sheites, twa pair of blankets, ane old silk quilted mat being of ane isaballa collour, with a coverlett conforme to the hingings; in the draw bed, a fethir bed, a bolster, a pair of blankets and a whyt worset rug; a timber table; thrie pots; and a drystooll. All the roumes and chamberes within the house, high and laigh, are compleitlie furnished in doores, lockcase, windowes and casements conforme, and the stables all in order, with heck and manger.

R. Mr. of Herreis.

J. Maxwell, witnes. Alexander Maxwell, witnes.

R. Gordoune, witnes. W. Gordoune of Penninghame witnes.

Upon the secund of March, taiken to the Newtoun out of this within wryten inventir, for the accomadation of my lord's funerall, to wit, aughtein

them into England, he was taken prisoner at Preston, and was beheaded on Towerhill, on 24th February, 1716.—His titles and estates were forfeited to the crown. His eldest son,

Robert Gordon of Kenmure died in 1741, and was succeeded by his brother,

John, an officer in the army, who, dying in 1769, was succeeded by his second but eldest surviving son,

John, a Captain in the 17th regiment of foot, who, upon his grandfather's forfeiture being reversed, in 1824, was restored to the dignity of Viscount Kenmure, &c. He dying in 1840, was succeeded by his nephew,

Adam Gordon, an officer in the Navy, since whose death, in 1847, the title has been unclaimed.

Gordon of Earlston.

The family of Earlston were amongst the first in Scot land to adopt the opinions of the Reformation and were zealous in maintaining the Presbyterian form of worship, during all the attempts which were made to establish Episcopacy as the national form of church government in Scotland. It is said that when some of the followers of John Wickliffe, fleeing from persecution, took refuge

chayres coured reed Rusew leathir; with twelff pewder plaits and my lord's armes thairupon, with sax uther plaits; four candle sticks with my lord's armes upon thame, of pewder; and two ordinarie candle sticks. Item, the haill caice of kniffes conforme to the within wryten inventir. Item, thair is over and above quhat is wrytten in the inventir on the bak syde, left in the parlour, ane large keicking glass mulered with eibonie and caice conforme; ane uther keicking glass indented with the mother of pearls.

in the wilds of the Glenkens, they were entertained by Alexander Gordon, the laird of Earlston, who had a New Testament in the vulgar tongue which they used to read in their meetings in the wood near Airds.

Shortly after Sideserfe had been appointed Bishop of Galloway, in 1634, Alexander Gordon of Earlston having opposed the settlement of a minister who was very unacceptable to the inhabitants of the parish, was summoned by the bishop to appear before the diocesan Commission Court and failing to obey was fined and banished to Montrose. He afterwards took an active part with the Covenanters, was at the Assembly of 1638, and was one of the Commissioners for the Stewartry in the Parliament of 1641. He is said, in Livingstone's Memoirs, to have been a man of great spirit and a worthy Christian, and to have refused to be made a knight when that honour was offered to him.

In 1663, when a curate was presented by the bishop as qualified for the charge of the parish of Dalry, William Gordon of Earlston was required, as being patron and a person of influence in the parish, to countenance his induction. With this injunction, after stating his reasons, he refused to comply, and was cited before the Council to answer for his contumacy. The Council being a short time after that informed that he kept conventicles and private meetings at his house, and had also attended several conventicles in Corsack wood and the wood of Airds, at which Mr Gabriel Semple, the ousted minister of Kirkpatrick-Durham, (most probably a brother of John Semple of Carsphairn,) had officiated, passed a most rigorous act against him, in which he is commanded to be banished from the kingdom within a month, not to return under pain of death, and that he live peaceably during that time under the pain of ten thousand pounds, or otherways to enter his person in prison. It would however appear that he did not obey this sentence, as in 1667, after enduring many hardships, he was by Bannatyne and his party of troops turned out of his house, which was then made a garrison. He subsequently suffered much persecution, and at length, in 1579, having sent his son forward to the army of the covenanters at Bothwell Bridge, he was hastening forward himself to their assistance, not knowing of their defeat, when he was met by a party of English dragoons and refusing to surrender was immediately killed.[1]

Alexander Gordon, his son, who was in the action at Bothwell Bridge, upon the defeat of the covenanters, fled from the field and escaped by the ingenuity of one of his tenants, who, recognising him as he was pursued through Hamilton, made him dismount and concealing his horse's furniture, dressed him in women's clothes and sent him to rock the cradle. He was declared outlawed and fled into Holland; but having returned, he was apprehended at Newcastle, in 1683, on board a ship bound for Holland, and sent to Newgate, from whence he was taken to Scotland, where, after he had been several times examined by the Council, he was sentenced to be executed at the Cross of Edinburgh, on the 28th September. This sentence was not however carried into execution, but, after having been put to the torture, he was, by the influence of the Duke of Gordon, reprived from time to time, and at last was sent to the Bass, where he remained until he was released at the Revolution.

1 This gentleman's Bible and Sword have been preserved by the family and are now in the possession of William Gordon, Esq. of Greenlaw.

Maclellan, Lord Kirkcudbright.

The sirname of Maclellan is one of the most ancient in the South of Scotland. The family are considered to have come originally from Ireland or the Western Isles and settled in Balmaclellan, conferring their name upon the parish,[1] from whence they spread over Galloway, where the clan became so numerous that fourteen knights of the name are said to have existed in the district at the same time.[2] The Maclellans were anciently sheriffs of Galloway and the office continued long in the family.

When Douglas Lord of Galloway, in the reign of James II., formed an offensive and defensive league with the Earls of Crawford and Ross, by which he trusted to make himself so powerful as to be enabled to set the king's authority at defiance, he endeavoured to persuade Sir Patrick Maclellan of Bombie to join his party. Maclellan, although his lands lay in the center of Douglas' possessions in Galloway and were liable to be overrun and pillaged by his retainers, altogether refused to enter into any engagement with him; upon which the Earl, highly irritated at his

1 Barscobe, in Balmaclellan, appears to have been the first place of their residence in Galloway, and it is somewhat singular that, notwithstanding the numerous branches which the family was divided into and the various lands which were in their possession, Barscobe was also the last place which remained in the family and only passed out of their hands about seventy years ago.

2 According to tradition, Barscobe, Gelston, Borgue, Troquhain, Barholm, Kirkconnel, Kirkcormock, Colvend, Kirkgunzeon, Glenshinnock, Ravenston, Kilcruickie, Bardrockwood, and Sorbie, were the places of which the various knights were proprietors and from which they took their titles.

refusal, collecting a number of his followers, surprised the Castle of Raeberry and taking Maclellan prisioner carried him to the Thrieve. Sir Patrick Gray, commander of the royal guard, who was Maclellan's maternal uncle, being apprized of this outrage, applied to the king in behalf of his nephew, and having obtained a letter requesting Douglas to deliver up his prisoner to him, immediately proceeded to Thrieve Castle, where he arrived just as the family rose from dinner. Douglas, who suspected the purport of his visit, declined entering upon business until Gray should have dined, saying, "it's ill speaking between a fu' man and a fasting ane," and privately gave orders that Maclellan should be beheaded. When Sir Patrick had dined, he presented the king's letter to Douglas, who, pretending to receive and read it with the utmost respect, took him by the hand and leading him to the place where the headless body of his nephew lay, said, "here is your sister's son, though he wants the head, his body is at your disposal." Gray, suppressing his indignation, replied, "no my lord, since you have taken his head you may take his body also;" and calling for his horse, he immediately mounted, and as soon as he had cleared the draw-bridge, upbraided Douglas as a blood-thirsty coward and a disgrace to knighthood, adding, "if I live you shall dearly pay for this day's work. Nor was it long till he had the opportunity of revenging his kinsman's death; for shortly after that Douglas was invited to visit the Court, under letters of protection, and after supping at the royal table, the king expostulated with him upon the evil tendency of such a bond as that which he had entered into and urged him to abandon it; this Douglas haughtily refused to do, and an angry altercation ensuing, the king at length became so enraged that drawing his dagger he exclaimed, "by Heaven my lord, if you will not break the bond this will," and immediately plunged it into the Earl's throat; upon which Sir Patrick Gray, who was keeping guard in a neighbouring chamber, immediately rushed in and dispatched him by a blow on the head with his pole-axe.

According to Crawford, the death of Sir Patrick Maclellan was so deeply resented by his relations and friends, that they committed great depredations upon Douglas's lands in Galloway, without having any warrant or authority to do so; for which the Laird of Bombie and most of his friends were forfaulted. The Barony of Bombie was recovered in the same reign by young Maclellan, who learning that the King had offered it as a reward to any person who should disperse a company of gipsies, which then infested Galloway, and bring their captain dead or alive to him, assembled some of his friends and having defeated the gang and killed their leader, he carried the head of the gipsey chief to the court and presented it to the king on the point of his sword, whereupon he was immediately seized in the Barony of Bombie; and to perpetuate the memory of the deed took for his crest a Moor's head on the point of a sword, and Think on, for his motto.

After the fall of the Douglases, Kirkcudbright, which had only been a burgh of regality under their jurisdiction, was created a royal burgh by a charter dated 26th October 1453, and the Maclellans had long the principal direction of affairs in the burgh and held the office of chief magistrate.

In May 1471, a charter of Lochfergus and other lands was granted to William Maclellan of Bombie, and his son and successor, Thomas Maclellan, had charters of various lands in the Stewartry between the years 1490 and 1500. Thomas, who died in 1504, was married to Agnes, daughter of Sir James Dunbar of Mochrum, by whom he had three sons,

1st., Sir William Maclellan of Bombie;

2nd., Gilbert, ancestor of the late Lord Kirkcudbright;

3rd., John Maclellan of Auchlean, whose male line becoming extinct, his estates returned to the family.

Amongst the many gentlemen connected with Galloway who fell at Flodden, was Sir William Maclellan of Bombie. He was considered to be one of the most accomplished gentlemen in Scotland, and had joined the Scottish army with a large number of his kinsmen and dependents—few or almost none of whom returned from that disastrous field. His son Thomas Maclellan of Bombie was killed on the High Street of Edinburgh on 11th July, 1526, by the Barons of Drumlanrig and Lochinvar with whom he had a feud. For this murder letters of Slaines were granted by Thomas Maclellan, the son of the deceased gentleman, in 1544, to James Gordon of Lochinvar and his assisters.

In 1562, Sir Thomas Maclellan of Bombie obtained a charter of the ground on which the church and buildings of the Grey Friers of Kirkcudbright had been constructed, with the orchards and gardens belonging thereto. He afterwards, in 1570, disposed of the Friers' Kirk and St Andrews Kirk to the Burgh of Kirkcudbright for two hundred merks, Scots, and ane hundred bolls of lime,—most probably the lime was to be used in the building of his Castle in Kirkcudbright which was finished about 1582.[1] He was married to Helen, daughter of Sir James

1 The Friar's Kirk was then used as the parish church, and St Andrew's Kirk having ceased to be employed as a place of worship was converted into the tolbooth of the burgh. In the disposition made by Sir Thomas Maclellan to the burgh, he binds himself to uphold the queyr, that is the east part of the said Friar's Kirk, and also to compel the parishoners to uphold the other two thirds of the same in thack, tymmer and stanes, as it was then delivered to the burgh. The east part of the kirk is still standing and formed part of the parish church until the erection of a new

Gordon of Lochinvar, by whom he had a son, Sir Thomas Maclellan of Bombie who died in 1597,[1] leaving three sons,

1st., Sir Robert Maclellan of Bombie,

2nd., William, of Glenschannoch,

3rd., John, of Borgue.

Sir Robert Maclellan of Bombie was made a knight by James VI., and appointed to be one of the gentlemen of the king's bed-chamber; in which office he was continued

one in 1838;—it is now used as a school room. Below it is the family vault or burying place of the Maclellans, over which is a monument, with the full length figure of a knight in complete armour reclining thereon, bearing the following inscription:—

IDOMINVS . SITVS . EST . T . M'LELLANVS . ET . VXOR .
D . GRISSEL . MAXVELL . MARMOR . VTRVMQVE . TEGIT .
HIS . GENITVS . R . D . KIRKCVDBRIVS . ECCE . SEPULCHRVM .
POSVIT . HOC . CHARI . PATRIS . HONORE . SVI .
ILLE . OBIIT . ANNO . DOM . 1597.
RESPICE . FINEM . MEMENTO . MORI . MORS . MIHI . VITA . EST .

1 It would appear that Sir Thomas Maclellan was also a considerable proprietor in Wigtown. We have annexed an extract from a contract, dated at Kirkcudbright, 15th December, 1586, by which Thomas Maclellan of Bombie receives the lands of Auchenflor, near Kirkcudbright, from Roger Kirkpatrick in Dursdow, in exchange for various houses and pieces of land in Wigtown. "The said Thomas grants him to haif sauld and heritabillie disponit, lykas be the tenor heirof he sells and heritabillie dispones to the said Roger Kirkpatrick, his aires and assignayes, in excambition with the said Roger's part of the landis of Auchenflor foirsaid, all and haile the said Thomas' tenements, landis, annualls, and pertinents underwrytten, lyane within the burgh of Wigtoun and territorie of the samen; togidder with all ryght, clame, interest, propertie and possession he, his aires and assignayes, had, hes or may haif thairto in tyme cuming, viz:—ane aker of beirland lyane on the west end of the burgh of Wigtown, beside the hous callit Roam's hous, betwixt the akeris of Patrick Blane on baith the sides. Item, ane medow adjacent to the said aker and utheris akeris lyane besyde the samen aker. Item, ane aker of beirland lyane betwixt the said landis and the beirlandis of Maielandis and the beirlandis of umqle the Freires of Wigtoun. Item, ane annual rent of twentie shillings money of Scotland, yeirlie, to be uplifted out of the tenement and yaird of umqle Johne Muirheid, lyane in the middes of the king's street on the eist pairt of the said burgh. Item, ane tenement and yaird lyane on the north pairt

by Charles I., who raised him to the rank of Baronet; and by letters patent, dated 26th May 1633, he was preferred to the peerage with the title of Lord Kirkcudbright. He died 1640 and the title devolved upon his nephew,

Thomas, (2d Lord Kirkcudbright,) the son of William

---

of said burgh, betwixt the tenement of the umqle Finlat Campbell of Corswell on the west, and the tenement of the umqle Johne Carsane on the eist. Item ane tenement sumtyme perteining to Michell Macclellane with front and yaird thairof, lyane on the south pairt of the said burgh, betwixt the said Thomas' uther tenement on the west, and the tenement of the umqle Johne M'Culloch in Kirkdale, inhabited by Johne Auld, on the eist. Item, ane tenement with yaird and barne thairof, lyane on the south side of the said burgh, betwixt the tenement of the umqle William Tait on the west, and the tenement of the said umqle Michell Maclellan on the eist. Item, ane tenement, callit the Ladie Bardochwode's tenement, lyane on the north pairt of the said burgh, betwixt the tenement of Johne A'Hannay on the west, and the said Johne A'Hannay's uther tenement, callit the new bigging, on the eist. Item, ane annual rent of sixtein shillinges money yeirlie, to be uplifted out of the tenement of the umqle George Inglis, now perteining to Morioun Inglis and William Dalzell, her spous, lyane on the north pairt of the said burgh, betwixt the tenement of William Gordoun on the west, and the tenement of the umqle Gilbert Clugstoun on the eist, Item, ane uther waist tenement adjacent to Sanct Bryde's well betwixt the common venall that passes to the said well on the eist, and the tenement of the umqle Patrick M'Kie, now perteining Duncan Warlaw on the eist pairtis. Item, ane croft of land callit the Lochane moir croft, lyane on the south pairt of the said burgh of Wigtown betwixt the lands of Johne M'Ilhanche on the south and the twa common king's vennales on the ane and uther pairtes, and ane oxgang land of the common lands of the said burgh, perteining to the said Thomas in rentall. And for the said Roger's better securitie thairof, the said Thomas Maclellan of Bombie bindes and obleisses him, his aires and assignayes, to infeft and seis sufficientlie, be charter and seasing, in dew forme, the said Roger Kirkpatrick, his aires and assignayes, heritabillie, in all and haill the aikairs land, in dew tenements, annuel rents, and croft of Lochane moir, foirsaid, to be haulden of the said Proveist and bailleis of Wigtoun, supperiors thairof, be resignation, or to be haulden of the said Thomas and his airis in blanche forme, be peyment to the said Thomas and his aires of ane penny yeirlie, at witsonday, gif it beis requyrit, and to the town of Wigtoun the mailles and dewteis thairof, usit and wont, conteinit in the said Thomas infeftment thairof.

Maclellan of Glenshannoch. This nobleman was a zealous presbyterian and took a warm interest in all the affairs of the Covenanters, from the very commencement of their differences with the King. In 1639, he was at the Scottish Camp at Dunse Law with a considerable number of troops, and, when articles of pacification had been entered into between the king and the Scottish Commissioners and an end put to the further progress of hostilities at that time, he attended the Assembly of 1639, when, along with several ministers and noblemen, he was appointed to examine a book called the Large Declaration. This work, which was written by Dr Balcanquel, was violently opposed to the principles of the covenanters and the lawfulness of the covenant, the adherents of which it represented in a most gross and iutemperate manner.—The reports which were given in to the Assembly, by those who had been appointed to examine the book were highly condemnatory; and upon the Moderator desiring that some of the brethern should give their opinions of the work, Mr Andrew Cant said, "It is so full of gross absurdities that I think hanging of the author should prevent all other censures;"—the Sheriff of Teviotdale was of the same opinion, saying, "I could execute the sentence with all my heart, because it is more proper for me, and I am better acquainted with hanging;"—and Lord Kirkcudbright, referring to the various persons whose ears had been cut off in England for writing against Laud and the prelates, said, "it is a great pity that honest men in Christendom, for writing little books called pamphlets should want ears, and false knaves for writing such volumes should brook heads."

Lord Kirkcudbright was in 1640 appointed Colonel of the South Regiment, a body of troops which was raised, principally in Galloway, to assist the Covenanters, and accompanied the army into England. He attended the Scottish Parliament in 1644; and the office of Steward of the Stewartry of Kirkcudbright, having become vacant by the forfeiture of the Earl of Nithsdale and his deputies, he was by the Parliament created Steward of that Stewartry and a commission was granted him to hold that office until the next triennial Parliament.

In 1645 a regiment of foot which Lord Kirkcudbright had raised at his own expense, principally from among his own vassals and retainers, having behaved with extreme gallantry at the Battle of Philiphaugh, the Scottish Parliament, as a reward for the good services which they had then rendered awarded them 15000 marks out of the forfeited estates of Lord Herries.

Thomas, Lord Kirkcudbright, died in 1647 and, leaving no issue by his wife Lady Janet Douglas, second daughter of William first, Earl of Qeensberry, the title and estates divolved upon his cousin,

John, (3rd Lord Kirkcudbright,) the eldest son of John Maclellan of Borgue. He like his predecessor was an enthusiastic presbyterian.

Immediately after the defeat of the forces which had been sent into England by the Scottish Parliament, in 1548, for the relief of Charles I, the ultra-covenanters or Kirk party, of which Lord Kirkcudbright was a supporter, and who had received the intelligence of the rout with the feelings of the highest exultation and looked upon it as a manifestation of the divine wrath against the engagement which had been entered into by the Scottish Parliament in favours of the king, and as a heaven-sent recognisance of the justness of the covenant, stirred up an insurrection, and Lord Kirkcudbright raised a body of levies to join the Whigamore Raid, as it was called. This party, having fixed their head quarters at

Edinburgh, in a short time became the most powerful in the kingdom and constituted themselves into a Committee of Estates. In September they sent Commissioners to Cromwell, whose quarters were at Berwick, soliciting him to join the honest party at Edinburgh, as they termed themselves. This invitation was cordially accepted by Cromwell, and Lord Kirkcudbright and General Holburn, having been appointed as a deputation from the Committie of Estates, met him at Seaton who accompanied him to Edinburgh.

The Amity of the Scottish and English Parliament was however of very short duration. for, as soon as the tidings of the execution of Charles I. reached Edinburgh, the Estates, feeling indignant that their remonstrances in behalf of their sovereign had been disregarded by the English Parliament, passed an Act on the 5th of February, 1649, for proclaiming his eldest son, Charles, as king, and ere long, were again embroiled in a civil war. Lord Kirkcudbright's regiment, which had been sent to Ireland, was, on the 6th December 1649, attacked by the Parliamentary Forces at Lessnegarvey, in Ulster, and being defeated were nearly cut to pieces.

Although the estates which this Lord Kirkcudbright succeded to were almost of a princely extent, he expended such large sums in the furnishing of the various forces which he raised during the Civil Wars, and for which he never received any remuneration, that he was reduced to a state of comparative indigence. After the Restoration his estate was altogether ruined by his opposing the introduction of a curate into the church of Kirkcudbright, in 1663. He died in 1664, leaving by his wife Anne, daughter of Sir Robert Maxwell of Orchardton, a son,

William, (4th Lord Kirkcudbright,) who died under age and without issue in 1669. The whole estate having been carried off in his minority, at the instance of his father's creditors, there was nothing left to support the title when the succession opened upon his cousin-german,

John, eldest son of William Maclellan of Auchlean the second son of John Maclellan of Borgue. This gentleman was entitled to have succeeded to the honours and dignities of Lord Kirkcudbright but appears never to have assumed the title. He died under age and without issue and his brother,

James Maclellan of Auchlean, (5th Lord Kirkcudbright,) who did not assume the title till the keenly contested struggle for a representation of the Scottish peerage betwixt the Earls of Eglinton and Aberdeen in 1721.—He was served nearest and lawful heir male to his uncle, John, Lord Kirkcudbright, on 15th February 1729, and died in 1730, leaving no male issue, when the title devolved upon

William Maclellan of Borness, (6th Lord Kirkcudbright,) the heir male of the body of Gilbert Maclellan, the brother of Sir William Maclellan of Bombie who fell at Flodden, He died about 1762 and was succeeded by his eldest surviving son,

John, (7th Lord Kirkcudbright,) who was an officer in the 30th regiment of foot, in which he had the commission of ensign in 1756 and that of lieutenant in 1758. He had a company in the 30th regiment of foot in 1774, and exchanged it for a lieutenancy of the 3rd regiment of foot guards in 1776, in which regiment he had a company with the rank of lieutenant colonel in 1784, and retired from the service in 1789. His Lordship died at London in December 1801, and was succeeded by his eldest son,

Sholto Henry, (8th Lord Kirkcudbright,) who died in 1827, and was succeeded by his brother,

Camden Grey, (9th Lord Kirkcudbright,) an officer in the Coldstream regiment of foot guards, in which he had the commission of an ensign in 1792, that of lieutenant in 1794. He quitted the service in 1803, and died at Bruges, in Belgium, on the 19th April 1832. Upon his death the title became dormant.

John Fullarton of Carleton.

John Fullarton of Carleton, who is mentioned in Livingston's Memorable Characteristics as a grave and cheerful christian, was the representative of a family who at one time possessed considerable lands in Galloway.[1] He appears to have been the author of some poetical pieces which were published and circulated amongst the cove-

---

1 In 1586, John Fullarton of Carleton granted security on his lands of Cotland in the parish of Wigtown, to Hellen Fullarton, relict of David Gordon in Laggan, and William Gordon her son, for payment of two hundred merks which he had received in their name from Sir John Gordon of Lochinvar, for redemption from them of the twa merk lands of Killegown which they had in alienation of Thomas M'Culloch of Cardiness

The lands of Littleton in Borgue were wadset to William Fullarton, brother german of John Fullarton of Carleton, by George Muirheid and his mother Marion Reddick in 1586.—(Register of Deeds of the Stewartry of Kirkcudbright.)

In 1715 when Dumfries was threatened with an attacked by the jacobite forces under Lord Kenmure, Captain Hugh Fullarton, late provost, Samuel Ewart and Sergeant Currie commanded a body of foot who were sent from Kirkcudbright to the assistance of that town.

nanters during the Civil Wars. He also wrote a small religious volume, partly in prose and partly in verse, entitled, "The Turtle Dove, under the Absence and Presence of her only Choise," which was published at Edinburgh in 1664. This volume was dedicated by the author to Lady Jean Campbell, Viscountess Kenmure, with whose family he was connected, and is now so extremely rare that the only copy known of is in the hands of the publisher. We have therefore been induced to give a few extracts from it, and a copy of a poem on the causes of the covenant.

(Extract from the Introduction to the Turtle Dove)

"Amongst many other inumerable testimonies of Gods unspeakable goodness (when contrary to all appearance or probability of expectation) there was by a gracious dispensation offered unto me, the liberty, freedom and benefite of a long wished for retirement, I took the opportunity of the time, to take notice of the most observable passages of 72 years time passed me: Wherein for the space of 36 years from my birth, being carried by the swey of mine own naturall and native inclination, the concourse of the like company, and the current of time and place.

1. First, in my minority I have had much fearful proof of what impiety the blind born, sin-born, wrath-born, brutish atheist is naturally poysoned with, and prone unto.

2. And secondly, after the years of discretion, and the benefite of breeding a more fearful proof of the vanity and deceit of self confidence, upon the ground of civile carriage and commerce.

3. Thirdly, and most dreadful of all, the conviction of that villany hypocrisie, under a formality of profession.

After all, it being the good pleasure of my God, in a time of love, by the Ministry of the Word, and convoy of the Spirit, to clear mine eyes, and open mine ears, unto a right uptaking of mine own by-past madnesse and misery, and of His marvelous long sufferings and mercy, so as (by grace) might have both humbled me, and helped me.

And now again, for the space of other 36 years, being under the name of a standing (but more properly stammering) Professor, I have found by better (althogh bitter) experience, that the way of Gods Children through the wildernesse is strawed with inumerable piercing thorns of divers afflictions and variety of temptations; And that the most searching tryals and sharpest afflictions are so unseparably conjoined, as fire and heat, under the exercise of desertion, being unto the spirituall [man and renewed party a present (but temporary) hell, and yet carrying a heaven in the heart of it."

An Acrostick upon the name of that very Religious and Famous Gentle-Woman, MARION M'KNAICHT.1

M More happy then imagined can be,
A And blessed are such as with heart sincere,
R Resolve to cleave to Christ, to live and die
I In Him, with Him, and for Him to appear,
O O what transcendent glorie grows from grace!
N None but, no not the soul refined shall
Mc Make to appear, that Light, that Life, that peace,
K Known only to the pure Possessors all.
N Now, thou by grace art unto glory gone,
A And gained the Garland of eternall blesse,
I In seeing Him, who on the glorious Throne,
C Created, uncreated, glory is:
H Heavens Quire did sing at thy conversion sweet,
T Time pasts thy finall comforts to compleat.

The Manifesto of the Nobilitie, Gentrie, Pastours, Burgors, of the Kingdome of Scotland.—Declaring the cause of this new solemme League and Covenant.

Wee Earles, Lords, Barrons, Pastours, Burgers, Knights,
Attest the *Antient* of the days and nights;
And in his presence wee doe manifestize,
Who spyes, and pryes in worldly secrecies.

1 A great many of Rutherford's Letters are addressed to this lady, who was much celebrated for her piety. She was married to William Fullarton, provost of Kirkcudbright, and was the daughter of John M'Knaught of Kilquhanidy, who was slain at the Carlinwark, in 1612, by Thomas and John Maxwell Her mother Margaret Gordon was sister to Lord Kenmure.

That neither by respect nor greed of Gaine,
Nor any popular Capritious Braine,
Nor any private purpose or intent,
For to assist the *English Parliament*,
But in so farre as wee and they agree,
Sinceirlie in *Religions* Puritie,
And wee doe manifest, and wee proclaime,
Our Soveraignes Honour justlie to maintaine,
Nor is it neglect of our Monarchs Throne,
Hath caused us to Re-covenate in one;
But a true zeale and a true reverence,
To Gods true worship, and our Faiths defence:
And we protest that wee doe love and tender,
(Nixt to our God,) King CHARLES our Faiths defender;
Yea, for His sake against all humane Lawes,
We will adventure in his Royall Cause;
Our honour, lives, estates, yea, more then this,
All to our Soules shall passe for Him and his.
But in the Cause of God, wee doe entreat
(As Supplicants, all prostrat at His Feete)
To grant us Audience, that wee may relate
Our Grievances and our deplored Estate;
And first of all wee plainlie manifest,
(Our God and all his Angels wee attest)
That our designment and intendement,
Is onlie for to make Attonement,
Betwixt our dear King and his Parliament,
This is the Scope of our new Covenant,
So this great mistrie which we have in hand
It doth surpasse the love of life and land:
O! who darre bee so bold to winke or nod,
Or play like bairnes at *Belliblinde* with God,
Religion daughter of our heavenly King,
Of mans Salvation the first Source and Spring,
Gods darling and the onlie linke and Chaine
That draws lost Soules from hells eternall paine,
To make a mock of *Thee*, or to dissemble
O! *All-beholding God*, for feare wee tremble,
Then for Thy sake, *Religion*, wee doe sweare,
To byde all straits Humanitie can beare;
For, for this Cause which wee doe now maintaine,
Esay in pieces cut was hasht and slaine.

Ierome was stoned, and *Daniel* was indened,
Amos was rent, *Paul* by the sword did end.
And sith all those, most voide of any feares,
Faught in this Cause before us many yeares,
Shall wee turne backe, or like base cowards yeeld,
To hazard stoutly in that glorions field;
Wherein our Saviour as great *Generall*,
The Prophets, and the blessed Apostles all,
Lyes in the bed of Honour? No, wee scorne
To retrograd like femall Culyeons borne.
Let timorous slaves bee servill to their feares,
Our Cause amounts our hearts above the Sphears.
Death in a clap freeth men from sin and errour
To Spirits of highest ranke, it is no terrour;
For our *Divines* hath catechiz'd us so,
That Sathans Hang-man Heresie, wee knoe:
To bee a totall destitution
Of joyes eternall, and Salvation,
A witching ghost, a cast-away of Grace,
A pull-backe from the presence of Gods Face,
A fearefull and a finall fall-away,
From comfort of the *Holy Ghost* for aye.
O! wee doe quacke, when that wee thinke upon
This head-aik, heart-aike, definition.
So wee resolve to conjoyne heart and hand,
From heresie, to purge this happie Land:
And to set up upon our highest Mountaine,
Gods Tabernacle in this Yle of Britaine,
That all the world may glorifie our God,
Who on our hills hath flourished *Aarons* rod.
The hearts of Kings are only in Gods hands,
Hee turnes it to his will, when hee commands.
God save our King, God turne his Royall Heart
From those, who doth from us His Minde divert.

Maclellan of Barscobe.—The Rising at Dalry.

On the 13th of November 1666, a small party of Sir James Turner's soldier's, who were then quartered in the Stewartry for the purpose of exacting the fines imposed upon the presbyterians, in the execution of which office they displayed a violence and brutality truly demonical, had seized upon some corn belonging to a poor old man of the name of Grier, in St John's Clauchan of Dalry, and were useing him in a most barbarous and inhuman manner. This being told to four countrymen, one of whom was Maclellan of Barscobe, who were taking some refreshment in a small tavern called the Midtown in Dalry, they resolved to use their utmost endeavours to release the aged sufferer, and proceeding to the house in which the soldiers were exercising their cruelty earnestly solicited them to forbear torturing the poor man. This however these ministers of oppression refused to do, and after some altercation drew their swords and severely wounded two of the countrymen, one of whom immediately discharged a pistol, loaded with a piece of a tobacco pipe instead of a bullet, at one of the soldiers, and an encounter followed, which ended in the defeat of the military.

The countrymen now saw that they would be acconnted as rebels, and fearing lest another company of about ten or a dozen soldiers, who were in another part of the same parish, should be apprised of what had occurred, they resolved to take precautionary measures to secure their own safety. Accordingly, the next day, with the assistance of some others who had joined their party, they surprised the soldiers, all of whom surrendered except one man who was killed.

On the 15th Mr Maclellan of Barscobe, Mr Neilson of Corsock and some other gentlemen collected about fifty horsemen and a small party on foot and proceeded to Dumfries, where they surprised Sir James Turner in his chamber and seized the money which had been sent from Edinburgh to pay the troops as well as that which was the proceeds of fines recently imposed. This money was entrusted to the care of a Stranger, calling himself Captain Gray, who had joined them the preceding evening, and decamped the following night carrying it all off with him. Thus commenced the rising of the covenanters which ended so disastrously at Rullion-green.

From the almost universal discontent which prevailed throughout the country it was thought the rising would be general. As soon as the news of the commotion reached Edinburgh, the presbyterians in that city met and deliberated upon the expediency of affording assistance to the insurgents, and Colonel Wallace Mr Welsh of Irongray and several others immediately set out and joined them at the Bridge of Doon in Ayrshire.

Colonel Wallace was chosen commander at Ochiltree, where they first assumed the appearance of an army and held their first council of war, at which it was resolved to march towards Edinburgh.

Upon Friday, the 23rd November, they marched to Cumnock, where they rested for a short time, and then proceeded to Muirkirk, the same evening, under a heavy rain, through a long and deep moor.

On Saturday they reached Douglas, where a Council of War was held, at which, having debated whether they should continue in arms or disperse, it was determined to proceed with their undertaking. A proposal was also made at this meeting to put Sir James Turner, whom they still carried with them as a prisoner, to death, but was rejected.[1]

On Sabbath they marched towards Lanark where the Covenant was renewed on Monday morning, and a considerable number joined the small army, so that when they left it in the afternoon, they amounted to about three thousand men. Entertaining some hopes that they would be joined by some in West Lothian and Edinburgh, they resolved to march east-ward and on Monday night, some hours after day light had disappeared, after a weary and toilsome march, through almost impassable moors, during which the rain fell in torrents, they reached Bathgate.—

1 Sir James Turner in his Memoirs, says,—"I was taken into a contrey house under pretence to refresh; but it was that I sould not looke upon thier armie (for so they were pleased to call it), till they had marshalld it rightlie. At length I was mounted and led along the reare of both horse and foot; and therafter I was brought to the front of the battell, where I did not let the opportunitie slip to reckon them. I found their horse did consist of foure hundreth and upwards, besides the partie of horse which was at Lainrick, and some other small parties which they had sent abroad to plunder horses,—a Sundayes exercise, proper onlie for phanaticks. The horsemen were armed for most part with suord and pistoll, some onlie with suords. The foot, with musket, pike, sith, forke, and suord; and some with staves, great and long. There I saw tuo of their troops skirmish against other tuo (for in foure troopes their cavallerie was divided), which I confess they did handsomelie, to my great admiration. I wonderd at the agilitie of both horse and rider, and to see them keepe troope so well, and how they had comd to that perfection in so short a time."

"That night a councell or committee was keepd where it was concluded, that nixt morning, the Covenant sould be renewd and suorne. And the question was, whether immediatlie after they sould put me to death, they who were for it pretended ane article of the Covenant obliged them to bring all malignants to condigne punishment. Bot it was resolvd that I sould not dy so soone, bot endeavors sould be used to gaine me. All this was told me by one of my intelligencers before tuo of the clocke nixt morning. Yet I have heard since, that it was formallie put to the vote whether I sould die presentlie or be delayed, and that delay was carried in the councell by one vote onlie."

As no accommodation could be procured for so many men, and as they had received intelligence of the near approach of General Dalziel, who had collected a considerable body of troops and followed fast in their rear with the design of attacking them, they resumed their march the same night, at twelve o'clock, and during this dreadful night nearly half of their number had disappeared. When they reached New-Bridge, on Tuesday morning, they presented the appearance of a wretched worn out crowd rather than an army.[1] They then proceeded to Collington, from whence, passing the east-end of the Pentland Hills, they marched to Rullion Green, where, on the 28th November, they were attacked by the forces under General Dalziel and utterly defeated,—about fifty were killed and a hundred and fifty taken prisoners.

1 The following derisive description of the forlorn appearance presented by the covenanters is extracted from "The Whigs' Supplication;" a poem, by Samuel Colvil.—Edinburgh, 1711:—

"Right well do I the time remember,
It was in *Januar* or December,
When I did see the out-law Whigs,
Lie stattered up and down the rigs.
Some had hoggers, some straw boots,
Some legs uncovered, some no coats;
Some had halberts, some had *durks*,
Some had crooked swords, like Turks;
Some had slings, and some had flails,
Knit with eel and oxen tails;
Some had spears, and some had pikes;
Some had spades which delved dykes.
Some had fiery peats for matches;
Some had guns, with rusty ratches;
Some had bows, but wanted arrows:
Some had pistols without marrows;
Some had the coulter of a plough;
Some scythes both men and horse to hough;
And some with a Lochaber ax,
Resolved to give *Dalzell* his paiks.

In this engagement Mr Maclellan of Barscobe commanded the small body of horsemen from Galloway, which was in the covenanting army. He was afterwards imprisoned for a long time, but was released upon taking the bond of peace, by doing which he greatly offended some of the more zealous of the covenanting party. The following notice of Mr Maclellan's death appears in Law's Memorials;—" Nov. 1683,—Some of these men of wild principles go into the house of Barscobe a gentleman in Galloway, who had been of a long time prisoner for joining with the men at Pentland hills, and got free upon his taking of the bond of peace (which thing incensed them) and strangles him in his own house."

Some had cross-bows; some were slingers;
Some had only knives and whingers;
But most of all, believe who lists;
Had nought to fight with but their fists.
They had no colours to display;
They wanted order and array;
Their officers and motion-teachers
Were very few, beside their preachers.
For martial music, every day
They used oft to sing and pray;
Which *hearts* them more, when danger comes,
Than others' trumpets and their drums.
With such provisions as they had,
They were so stout, or else so mad,
As to petition once again;
As if the issue proved in vain,
They were resolved, with one accord,
To fight the battles of the Lord."

Cardoness.

The Castle of Cardoness, which stands on an abrupt eminence overlooking Fleet Bay and distant about a mile and a half from Gatehouse, is one of the most ancient fortlaces in the Stewartry. The keep, which is a strong substantial building of a square form, still rears its entire though roofless walls amidst the remains of the various offices and smaller buildings with which it was formerly surrounded.

It is impossible now to ascertain when it was built, as history is altogether silent both with regard to the period of its erection and to the names of its first proprietors. Tradition, indeed, states that it first belonged to a family of the name of Caird and that the erection of the Castle having exhausted the resources of no less than three successive lairds, the fourth was compelled to carry heather on his own back from Glenquicken Moor to thatch his domicile. This Laird is said to have been a most profane and wicked man and being resolved to acquire riches, without any regard as to the means employed, he connected himself with some of the Border Freebooters and the tower of Cardoness soon became one of the noted strong-holds of these banditti in Galloway. Having been married to a lady for nearly twenty years, during which she had given birth to nine daughters but no sons, when her confinement next approached, he threatened to drown her and all the daughters in the Black Loch unless she should present him with an heir to his name and honours. Great was the anxiety of the poor lady when the time of her confinement drew nigh, as she well knew her lord was not a man to use idle threats, and great were the rejoicings in the Castle when she gave birth to a son. As it was in the midst of winter and the Black Loch was completely frozen over, the laird determined to hold a fete upon it on the Sunday and thither, accordingly, the whole family were conveyed on that day, except one daughter who refused to accompany the party. In the midst of their enjoyment the ice rent around them and all were plunged into one central gulf. Thus perished the whole family except the one daughter, who was afterwards married to a gentleman of the name of M'Culloch, a family of very ancient standing in Western Galloway.

Whether this was the manner in which the M'Cullochs acquired Cardoness cannot be determined, but this tradition has at least long been held as authentic by the family, and ample proof exists in history that they were proprietors of the estate at a very early date.

In 1587 William M'Culloch of Myretown, with consent of his wife Marie M'Culloch of Cardoness and her mother Katherine Gordon, wadset the ten mark lands of Auchenlarrie, in the parish of Anwoth, to Sir John Gordon of Lochinvar.

John Gordon, who was proprietor of Cardoness during the Civil Wars, appears to have been warmly attached to the presbyterian cause and to have taken an active part in all the measures which the covenanters took to forward their cause in the south. He was appointed captain of part of the forces raised in the Stewartry in 1640, and was lieutenant-colonel in 1644.

Mr Rutherford, when minister of Anwoth, resided at Bush o' Beel, an old baronial house near the Kirk of Anwoth, which belonged to the Gordons of Cardoness, and contracted an intimate friendship with the family, which seems to have been contined long after he left the parish as many of his letters are addressed to Mr Gordon and his lady.

In the latter part of the seventeenth century two claimnants appeared for the estate of Cardoness,—Sir Godfrey M'Culloch of Myretown, who held possession, and William Gordon who resided at Bush o' Beel. A bitter animosity existed betwixt these persons and, in October 1690, Sir Godfrey having been persuaded by some persons to go with them to Bush o' Beel and assist in relieving some cattle which had been poinded, met Gordon and discharged a loaded gun at him, by which he was so severly wounded that he expired the same night. Sir Godfrey immediatdly fled to England, where he remained for a considerable time, but, having returned to Scotland, was apprehended on a Sunday while attending public worship in a church in Edinburgh. It is said that a gentleman from Galloway who was present cried out at the end of the service, "Shut the doors, there is a murderer in the house." Sir Godfrey was tried for the murder of Gordon on the 16th February 1697, and beheaded at the Cross of Edinburgh on the 26th March following.

Miss Stewart of Castle Stewart, the neice of William Gordon, was married to Colonel William Maxwell, the son of the Rev Mr Maxwell, the presbyterian minister of Minnigaff who was ousted out of his charge at the introduction of the curates in 1661. Colonel Maxwell came over with King William as an officer at the Revolution and served in the army till 1706. In 1715 he was appointed Governor of Glasgow which office he held till the Rebellion was over. For the valuable services which he then rendered he received a present of plate from the cities of Edinburgh and Glasgow, ornamented with the arms of both cities, the greater portion of which is still in possession of his great-grandson Sir David Maxwell Bart., the present proprietor of Cardoness. After the suppression of the Rebellion Col. Maxwell retired into private life but lived to witness the defeat of the last attempt which was made to reinstate the Stuarts on the throne. He died at his seat at Bardarroch, in June 1752, at the very advanced age of 97.

Committee of Estates.

On the second of June 1640, the Scottish Parliament assembled at Edinburgh and although the king, a few days previous, had sent a letter to some of the Privy Council empowering them to prorogue it, they having been persuaded by Sir Thomas Hope that such an order was unformal, failed to execute it; and the Parliament sat down and proceeded to business, without a commissoner from the king. Having elected Lord Burleigh for their president, they determined that the lesser barons should take the place of the bishops, and after ratifying a number of acts, proceeded to choose a Committee of forty individuals, part of whom were noblemen, part gentlemen, and part burgesses, who should coninue to sit after the parliament rose, and should exercise sovereign power over all the affairs of the State.

The following list of the Committee of Estates is given Straloch's MS.:—

Noblemen:—the Earls of Rothes, Montrose, Cassilis, Wigton, Dumfermline, and Lothian; Lords Lindsay, Balmerino, Coupar, Burleigh, Napier, and Leven.

Lords of Session:—Dury, Craighall, and Scotstarnet.

Gentlemen:—Sir Thomas Nicholson of Carnoch, Lawer, Sir Patrick Hepburn of Wauchton, Sir David Home of Wedderburn, Sir George Stirling of Kier, Sir Patrick Murray of Elibank, Sir Patrick Hamilton of Little Preston, Sir William Cunningham of Caprinton, Sir William Douglas of Cavers, James Chalmers of Gadsgirth, Sir Thomas Hope of Kerse, Drummond of Riccarton, Forbes of Lesly, and Mr George Dundas of Manner.

Burgesses:—John Smith, burgess of Edinburgh; Thomas Paterson, tailor, and Richard Maxwell, saddler in Edinburgh; William Hamilton, burgess of Linlithgow; Mr Alexander Wedderburn, clerk of Dundee; George Porterfield, bailie of Glasgow; Hugh Kennedy, bailie of Air; John Ruthford, provost of Jedburgh; Mr Alexander Jaffray, burgess of Aberdeen, (or Mr William More, bailie of Aberdeen, in his absence;) James Sword, burgess of St. Andrews; and James Scott, burgess of Montrose.

Mr. John Maclellan.

From the circumstance of Mr Maclellan coming from Ireland shortly before he was appointed minister of Kirkcudbright it has generally been supposed that he was a native of that country; this conjecture however seems to have been made without any other foundation, and it is more than probable that he was a native of Kirkcudbright, as some entries occur in the Burgh Records by which it would appear that his father Michael Maclellan was long both an inhabitant and burgess of that burgh.

Mr Maclellan having received a liberal education went to Newton in Ireland, where he kept a school for some time at which several young men were educated who afterwards became distinguished members of the ministry. Having afterwards been tried by the presbyterian ministers in the County of Down, he was licensed and preached in their churches until among others he was deposed and excommunicated by the bishops.

In the spring of 1636 he resolved to accompany a party of presbyterians, amongst whom were Mr John Livingston, afterwards Minister of Stranraer, and Mr Robert Hamilton afterwards Minister of Ballantray, who proposed going to New England. After much toil and many delays the party, consisting of about a hundred and forty persons, sailed from Carrickfergus on the 9th September, in a small vessel of about a hundred and fifty tons, called the Eagle's Wing, which they had got built for that purpose. They were however detained by contrary winds for some time in Lochryan, and after having put out to sea, when about halfway between Ireland and Newfoundland, they encountered a tremendous hurrican in which the rudder was broken and the ship sprung a leak. They then resolved to return and arrived at Carrickfergus on the 3rd November.

After this, Mr Maclellan remained for some time in the counties of Tyron and Donegall, preaching at the private meetings of the presbyterians, but being pursued by orders of the bishops, he came over in disguise to Scotland.

In 1638 he was appointed minister of Kirkcudbright and attended the General Assembly which was held at Glasgow, in October, the same year.

In July 1639, application was made by Lord Kirkcudbright to the Magistrates, requesting them to allow the same sum, two hundred marks, for the vicarage tiends, to Mr Maclellan as they had previously paid to his predecessor Mr Glendonyng, but they would not consent to this, and alledged that they only paid Mr Glendonyng fifty marks for the vicarage tiends and that the other hundred and fifty marks were merely given to him as a token of their respect and out of their own good will, and therefore refused to allow Mr Maclellan any more than the fifty marks for the vicarage tiends.

Shortly after Mr Maclellan had been appointed minister of Kirkcudbright he displayed more than ordinary dilligence in preaching against the corruptions of the times and in rebuking those who infringed upon the discipline of the church, by which he offended several of his parishoners.[1]

In the assembly of 1640, Mr Henry Guthrie, then minister of Stirling, brought in a complaint against private meetings of persons at night for the purpose of reading the Scriptures and joining in prayer, which had become very general throughout the South and West of Scotland. He spoke with great vehemence against these meetings and several other ministers espoused his views and desired that they should be suppressed. One of the Commissioners from Galloway denounced Mr Samuel Rutherford, Mr John

1 In October 1639, Gilbert Reid having threatened to shoot the minister with a pair of bullets was committed to prison from which he made his escape, but having been captured he was again lodged in jail, under the charge of three men, to each of whom he was to pay ten shillings for every twenty four hours he remained in ward, and had also to fee a man to go to Edinburgh to advise what satisfaction the magistrates should take of him for breaking the prison.

In May 1642, Janet Creichton was sentenced to stand at the Kirk door from the time the bell began to ring till the text was given out, with a paper on her head declaring her fault which was speaking misrespectfully to the minister in the Kirk, while in the discharge of his office.

Livingston and Mr John Macllelan as having been the great encouragers of these meetings within these bounds; upon which Mr Maclellan craved that a committee might be appointed to enquire into the abuses complained of, and shortly afterwards an act was passed ordaining;—

"1st., That family worship be performed by those of one family only, and not of different families.

"2d., That reading prayers is lawful when none of the family can express themselves properly extempore.

"3d., That none be permitted to explain the Scriptures, but ministers and expectants approved of by the presbytery."

Mr Maclellan appears by this time to have gained the good will of the Magistrates of Kirkcudbright, as they gave an order to the owners of the town meadows to give the sixteenth coll of all the hay thereon to Marion Fleming,[1] the minister's wife, as a gratuity bestowed by the town upon their minister, during his absence at the Assembly, and shortly afterwards they met with Lord Kirkcudbright to consult with him about building a manse for Mr Maclellan.

A petition was presented to the General Assembly of 1642, subscribed by a number of presbyterians in the North of Ireland, in which after detailing the deplorable condition they were in through want of the ministry of the Gospel, they desired that some ministers, especially such as had been driven away from them by the persecution of the prelates, might be sent over either to reside amongst them or to preach through the country.

[1] Marion Fleming was the daughter of Bartholomew Fleming, merchant in Edinburgh. Mr John Livingston minister of Stranraer was married to her sister. Barbara Hamilton, her aunt, was married to John Mein, merchant in Edinburgh, and their son, Mr John Mein, was minister of Anwoth about the year 1645.

After considering the petition the Assembly determined not to appoint ministers to any charges in Ireland, but that a few should go in rotation to the North of Ireland to visit, encourage and instruct the scattered flocks of Christ and, if need be, to try and ordain such persons as should be found qualified for the ministry. Mr Robert Blair, Minister at St Andrews, and Mr James Hamilton, Minister at Dumfries, were nominated to go for the first four months; Mr Robert Ramsay, Minister at Glasgow, and Mr John Maclellan, Minister at Kirkcudbright, for the next four months; and Mr Robert Baillie Professor of Divinity in the University of Glasgow, and Mr John Livingston Minister of Stranraer, for the last four months.

Besides being regarded as a devout and indefatigueable minister of the Gospel, Mr Maclellan was allowed by many of the more zealous covenanters to have been possessed of the spirit of prophesy.

He wrote an Account of Galloway, in Latin, for Bleu's Atlas, by which he gained some celebrity; and a short time previous to his death, which took place in the beginning of 1650, he composed the following epitaph on himself:—

"Come stingless death, have o'er; lo! here's my pass,
In blood character'd by his hand who was,
And is, and shall be. Jordan cut thy stream,
Make channels dry. I bear my Father's name
Stampt on my brow. I am ravish'd with my crown;
I shine so bright, down with all glory, down,
That world can give. I see the peerless port,
The golden street, the blessed soul's resort;
The tree of life, floods gushing from the throne,
Call me to joys. Begone, short woes, begone
I lived to die, but now I die to live,
I do enjoy more than I did believe.
The promise me unto possession sends,
Faith in fruition, hope in having ends."

It would appear that Mr Maclellan left no family, as Elizabeth Maclellan who was married to Alexander Mulcan, merchant and burgess of Kirkcudbright, is mentioned in the Burgh Records as having been the sister and heir of Mr John Maclellan.

John Ewart.

It is altogether uncertain at what time the Ewarts first settled in the Stewartry. Andrew Ewart, in Grange, who is repeatedly mentioned in the Burgh Records and Register of Deeds of the Stewartry, between the years 1576 and 1591, is the first of the family of whom we can find any notice. He was appointed Treasurer to the Burgh of Kirkcudbright in October 1583.

John Ewart, in Grange, was admitted freeman of the Burgh in 1601,—he was chosen bailie in 1611 and filled the offices of councillor and bailie various times between that date and 1635. He appears to have become proprietor of the lands of Mullock shortly after he had been made bailie.

John Ewart, younger, of Mullock, was a member of the Town Council in 1611, and was bailie various times between that period and 1630, when he was appointed to go as Commissioner for the Burgh to Jedburgh and then to proceed to Edinburgh to pay the cheker [Exchequer]. He was chosen Provost in 1649 and repeatedly filled the office of chief magistrate. He was one of the Commissioners appointed in the Stewartry of Kirkcudbright for the collection of the imposts which were raised for the purpose of maintaining the public tranquillity and restoring the prerogatives of the crown, by the orders of the first Parliament of Charles II., after the Restoration.

John Ewart, the son of the foregoing, was bailie 1653, and, in August of the same year, was appointed Commissioner for counting with the Commissioners at Leith regarding the duties of the town and other affairs connected with the Burgh. His first entry into the council appears to have been about the year 1647.

In the month of May, 1663, when a riot took place in Kirkcudbright, caused by the introduction of a curate into the church, to which the inhabitants were greatly opposed, the Privy Council appointed a Committee, consisting of the Earls of Linlithgow, Galloway and Annandale, Lord Drumlanrig and Sir John Wauchop of Niddry, to proceed to Kirkcudbright, accompanied by an armed force, and to make enquiries into it. They accordingly met at Kirkcudbright, on the 23th May, and ordered Lord Kirkcudbright with John Ewart, John Carson of Senwick and several others to be carried prisoners to Edinburgh, some for being concerned in the riot and others for not using their endeavours to quell the disturbance. The charge brought against John Ewart was, that he had been chosen provost of the Burgh at the previous election but had refused to accept of the office and was therefore the chief cause of the disorganisation of the magistracy,—it was likewise alledged that he had declined to give his advice and assistance in reducing the tumult, on the grounds that he was not a councillor. On the 13th August, the Privy Council passed the following sentence upon him,—"Likeas the said Lords banish John Ewart forth of this realm for his offence, and ordain and command him forth of the same betwixt and this day twenty days, not to be seen therein at any time hereafter, without licence from his Majesty or the Council, at his highest peril." He however obtained a mitigation of his sentence. A new election of councillors and magistrates was ordered by the Privy Council at this time, and Mr William Ewart, a younger brother of John, was chosen provost.

John Ewart does not appear to have taken any very active part in the concerns of the town, from this time until the Revolution, when the presbyterian party were again reinstated in power, and he was then again chosen provost and represented the Burgh in King William's first Parliament. While fulfilling the duties of Commissioner he was allowed the sum of three pounds, Scots, per diem, besides some other extraordinary expenses which might be incurred by him in furthering the interest of the Burgh.

He appears to have died about 1697, for Sir Andrew Home is then mentioned in the Burgh Records as having been elected Commissioner to Parliament, in the room of the late John Ewart.

His eldest son, the Rev. Andrew Ewart, was the first minister of the parish of Kells after the Revolution, and his grandson, the Rev. John Ewart, was minister of Troqueer in 1743. The last named clergyman's eldest son, Joseph, was Minister Plenepotentiary for Great Britain at the Court of Berlin, and his second son, William, was a distinguished merchant in Liverpool and the father of William Ewart, Esq., M. P., for the Dumfries district of Burghs.

Threave Castle.—Earl of Nithsdale.

When the Lordship of Galloway was conferred upon Archibald, the Grim, Earl of Douglas, by David II, in 1369, he built the stronghold of Threave on the remains of an ancient fortlace which had been the residence of Alan the last native Lord of Galloway, and it is said that the stones of the Old Abbey of Glenlochar, which was dedicated to St Michael and had been destroyed during some of the inroads of the English into Galloway, in the time of Edward I., were employed in the erection of the Threave. The Castle, which stands on an islet of about twenty statute acres, formed by the Dee ten miles above the estuary of that river, is yet a stately pile nearly seventy feet in height. The dungeon, arsenal and larder occupied the lowest storey; the second floor was occupied by the soldiers, and the third contained the state apartments.—It was surrounded by a barbacan flanked at each angle by a circular tower and secured in front by a deep fosse and vallum. After passing the drawbridge the only entrance was by a door placed so high in the wall that the threshold was on a level with the second floor. The door was secured by a portcullis constructed in such a manner as to slide in a groove of solid stone.

As long as the Douglasses retained their power Threave continued to be the place of their residence in Galloway, and they always kept a strong body of their retainers in it.—indeed it is said that William, the 8th Earl of Douglas, had a retinue of no less than a thousand armed men in the Castle.

Upon the forfeiture of that family, in 1455, the Castle of Threave, with all the lands and customs pertaining thereto, reverted to the crown, "never to be settled or bestowed, either in fee or franc-tenure, upon any subject whatever, except by the solemn advice of the whole parliament." It was then garrisoned by the king's troops, and a Lardner Mart Cow, that is a fat Cow fit for killing and salting at Martimnas, for winter provision, was yearly lifted in each of the twenty eight parishes of the Stewartry for the victualling of the garrison.

The customs and firms of the Castle of Threave were afterwards settled by James III. on his queen, Margaret of Denmark, as part of her dowery, and in October, 1477, Robert, son of John, Lord Carlisle, obtained a grant of the office of Steward of the Stewartry of Kirkcudbright and Keeper of the Castle of Threave.

After the surrender of Berwick to the English, in 1482, the Earl of Angus lost his office of Steward of Kirkcudbright and his command of Threave.

Sir John Dunbar of Mochrum, in 1502, obtained a grant for nine years, of the office of Steward of Kirkcudbright and Keeper of Threave Castle, by which he acquired the twenty mark lands of the Grange of Threave, the fishings of the Dee and the Lardner Marts, for which he engaged to pay the king a yearly rent of £100 and to keep the garrison on his own charge.

In February 1515-16, the office of Steward of Kirkcudbright and Keeper of the Castle of Threave, with the usual perquisites, was conferred by Queen Margaret upon Robert Lord Maxwell, as tutor of her son, for the term of nineteen years, without payment of any mails or duty whatever. And his Lordship, in 1526, obtained a grant in fee form, to himself and his heirs, of these offices with their pertinents.

The sons of Lord Maxwell held this fortress when it was stormed and taken by the troops of the Regent and Cardinal Beaton, in 1545.

When Lord Maxwell, in 1587, attempted to raise a force in the south of Scotland to assist Philip of Spain in the invasion of England, James VI took active measures to suppress the rising, and having collected a considerable number of troops proceeded to Dum ries, where he almost surprised Lord Maxwell, who however made his escape; and next day the castles of Thrieve, Caerlavrock and Langholm surrendered and were occupied by the king's forces.

In 1620, Robert, 9th Lord Maxwell, being in great favour with the king was created Earl of Nithsdale, and when the Civil Wars commenced he espoused the royal cause, to which he adhered during all the troubles which ensued. He held the Castle of Thrieve for the king, in which he placed a garrison of about eighty men besides officers, the greater part of whom were of the name of Maxwell; these he armed, paid and victualled at his own expense, until his Majesty, unable to send him any assistance, directed him on 15th September, 1640, to make the best conditions he could for himself.[1] The Castle was accor-

1 "Charles R.

"Right trusty and well beloved cousin and councilor, we greet you well. Understanding by this bearer, that although you were agreed with those that have beleagured you in Carlaverock upon honourable terms, for your coming forth and rendering thereof, yet that those conditions are not valid until such time as they be ratified by those that have made themselves members of the great Committee in Edinburgh, and fearing that your enemies there will not give way to your comming forth upon such good terms, we are therefore graciously pleased, and by these presents do permit and give you leave to take such conditions as you can get, whereby the lives and liberties of yourself, your family, and those that are with you, may be preserved: and in case they should urge the surrender of our castle of Thrieve, which hitherto you have so well defended, (and we wish you were able to

dingly surrendered to the covenanters and immediately afterwards dismantled and rendered untenable. Shortly afterwards the Earl of Nithsdale's lands were sequestrated, and he was subsequently imprisoned and suffered many hardships for his attachment to the royal cause. When he found he could render no further service to the king, he retired to the Isle of Man, where he died in the end of the year 1647.

Robert, the son of the Earl of Nithsdale, who succeeded to the title at his father's death, was also a zealous royalist; when only a young man, he was imprisoned by the parliament but liberated in 1646, upon finding bail to appear when called upon.[1]

---

do so still,) our gracious pleasure is, that you do rather quit the same unto them; which, if so, the necessity require you to do on the best and most honourable terms you can, rather than hazard the safety of your own person, and those with you; and in such case this shall be your warrant and discharge Given at our court at York, the 15th day of September, in the sixteenth year of our reign, 1640,"—Grose's Antiquities of Scotland.

1 The following extract from the Burgh Records of Kirkcudbright will show how numerous the clan of Maxwell had become in the Stewartry and the south of Scotland, about this time.

"At Kirkcudbright, the sextene day of October, 1661, in presence of the Magistrates and Council convened for the time.

"The quhilk day, ane noble and potent erle, Robert erle of Nithisdaill, Lord Maxwell, Eskdaill and Carleill: James Maxwell of Breckinsyde: Capitane Edward Maxwell, brother to my Lord Herres; Sir James Murray of Caleton, knicht; Thomas Maxwell, laird of Gelstoune; John Maxwell of Cowhill; John Maxwell of Gribtoune; William Maxwell, his son; William Maxwell of Killilung; Robert Maxwell of Garneshalloch; John Maxwell of Conhaith; Hugh Maxwell of Blackbellie; Patrick Maxwell, younger of Sprinkell; Robert Maxwell of Castlemilk: James Maxwell, Tutor of Tinwald; John Maxwell, younger of Holme; John Maxwell, younger of Broomieholm; Robert Maxwell, son to Broomieholm; Major Alexander Maxwell of Balmangan; John Maxwell of Killberne; Edward Maxwell of Woodheid; William Maxwell, son to Cowhill; John Maxwell of Steilstoune; Harbert Maxwell of Carse; Hugh Maxwell of Wraithes; John Maxwell of Keltoune; James Lidderdale of Isle; John Grier of Barjarg;

William, the 5th and last Earl of Nithsdale, sold the fishings which pertained to Thrieve Castle, in 1704, but retained the fortress and the right of the Lardner Marts. As these had not been seized at his attainder, in 1716, a claim was made by the family, at the abolition of the heritable jurisdictions in Scotland, in the year 1747, for parting with the superiority of the lands of Threave Grange and the yearly supply of fat cattle levied for the support of the garrison, which however was not sustained.

In the beginning of the present century, Thrieve Castle was partially repaired under the superintendence of Sir Alexander Gordon, Stewart of the Stewartry, for the purpose of converting it into a barrack for the reception of French prisoners, but was never occupied.

John Lindsay of Wauchope; John Roome of Dalswenton; Charles Murray of Barnhowrie; John Hairstaines of Craigs; John Dalrymple of Waterside; Thomas Dalrymple, his brother; Alexander Agnew, younger of Killumpha; Mr John Corsane of Barndarroch; John Grierson of Dalskarithe; Alexander Kirkpatrick of Penmadie; John Maxwell, Tutor of Garneselloche; James Maxwell, his son; John Young of Gulliehill; Patrick Young of Auchenskeoche; James Young of Broomierig; Robert Young, Chirurgeone; John Roome, apothecarie; Alexander Wallace of Carell; James Corsane, son to Barndarroch; John Maxwell of Arkland; William Herries of Langtoune; John Broune of Muirheidstoune; Robert Maxwell of Langridding; John Cuitlar of Orraland; John Sturgeon of Torrorie; John Maclellan, son to my Lord Kirkcudbright; James Welsh, Agent; Thomas Broune, Thomas Smith, Robert Glassen, William Carleill, Charles Maxwell, James Edgar, William Maxwell, servitors to the said erle of Nithisdale; John Grier, servitor to Sir James Murray; John Miller, servitor to Major Maxwell; William Carsane of Netherriddick; John Maxwell, servitor, to Lord Herries; Robert M'Briar, younger of Newwork; William Batie, servitor to Springkell; David Herries, servitor to Edward Maxwell of Woodheid; William Gordon, younger of Crogo; William M'Colme, servitor to my Lord Nithisdaill; Robert Maxwell, son to Robert Maxwell of Trostan; and John Hutoune, son to Thomas Hutoune; were admitted created and resaived bugesses of this burgh, who promeisit their fidelitie thereto, and to be assistant to the town and magistrates their of in their lawful offiars."

Mr George Rutherford.

Nothing is known of the history of Mr George Rutherford previous to 1629, when he was, through the influence of provest Fullarton and some other gentlemen, intimate friends of his brother Mr Samuel Rutherford, appointed schoolmaster and reader in Kirkcudbright. The duties of both of these offices he discharged to the entire satisfaction of the magistrates, and indeed of the whole community, till the end of the year 1636. At that time the Bishop of Galloway suspended Mr Glendonyng, the minister of Kirkcudbright, and gave the charge of the parish to a Mr George Buchannan, who was deposed by the Assembly of 1639.—The inhabitants however being all warmly attached to Mr Glendonyng, prevailed on him to continue in the exercise of his ministerial duties and refused to attend the ministrations of the Bishop's minion. As Mr George Rutherford was a zealous non-conformist and a strong supporter of Mr Glendonyng, he was summoned before the High Commission, and commanded to resign his charge and remove from Kirkcudbright. He then retired to Ayrshire, where he resided for some time; but soon after Mr James Scott, the minister of Tongland, had been deposed by the General

1 August 27th 1639—"Mr James Scott, was deposed for his absence from his flocke 8 Sabboths together, sacrialedge, intromitting with penalties and contributions, disobedience to the Presbytrie, tableing, converseing with excommunicat Papists, and declyning the Generall Assemblie."—Records of the Kirk of Scotland.

The Mr John Scott mentioned in the following extract from Kirkton's History of the Church of Scotland, is most probably the same person.

Assembly of 1639, he was appointed to the charge of that parish. Here he appears to have remained until his death, the exact date of which is uncertain, but most likely it was not long before the Restoration.

Mr Rutherford seems to have acquired the lands of Argrennan and Grayscroft, in the parish of Tongland, most probably by marriage, as, in the year 1668, his three daughters, Marion, who was married to Robert Hutton of Newlands, Martha, who was afterwards the wife of John Black, merchant in Kirkcudbright, and Barbara, were retoured as his successors and heirs in these lands. According to tradition Mr George Rutherford had a son, called Samuel, who was drowned when a child in the well at Argrennan.

As it may be interesting to know the duties discharged by a schoolmaster and reader at that time, we have subjoined a copy of Mr Rutherfurd's engagement with the magistrates when he entered upon these offices in Kirkcudbright.

"The Church of Ancrum being vacant through good Mr John Livingston's exile, one Mr John Scot, who hade been excommunicate 20 years before and still continued so, though he was possessed of two benefices elsewhere, is presented to the church Upon the day appointed for his induction, a number of the poor people conveened to give him the welcome abhorred pastors use to get, one poor woman amongst them desired earnestly to speak with him, hopeing to disswade him, but he fleing away from her, whereupon she takeing hold of his cloak to detain him, he beat her with a staff; this made some boyes throw a stone or two, which neither touched him nor any of his company; this is proclaimed a treasonable tumult. The countrey bailiffs both imprisoned and fyned them. But this saved them not from the claws of the commission, whither when they were brought, first four boys appeared. The Commissioner told them hanging was too little for them, so their sentence was to be scourged through Edinburgh, burnt in the face, and sold in Barbadoes. . . Thereafter they called for two brothers and their sister, the Trumbles of Astneburn; the two brothers, though both heads of families, they sent to Virginia, (to Barbadoes they would not; lest they should have the comfort of their neighbours) and the poor sister, a married woman, they appointed to be scourged through Jedburgh streets."

"At Kirkcudbright, the aught day of December, the year of God J^{m} vj^{c} twentie nine years.

"The quhilk day the proveist, baillies, counsell, and communitie of this burgh of Kirkcudbright, hes conduced and agriet with Mr George Rutherfurd, brother to Mr Samuel Rutherfurd, minister at Anwoth, to be their schuilmaister, within the said burgh, for the space of ane yeir, his entrie thereto beginning at Candlemas 1630 next; at quhilk tyme the said Mr George obleises him to enter to the said schuil, or within four or fyve dayes thereafter; and during the said space, to teich his scollers their learning, as they are able and capable to drink in the samen. Lykeas, the said Mr George obleises him, during the said space, to reid in the kirk the prayers, publicklie, morning and evening, to raise the psalmes in the kirk, publicklie, and to wryte in the kirk session, their acets, as he shall be requyred. For the quhilk the toun of Kirkcudbright obleises them to pay to the said Mr George, for keeping of the said schuil foure score punds, yeirlie, at four times in the yeir, Whitsonday, Lambmes, Hallowmes and Candlemes; and that everie toun bairn shall give him quarterlie viijs., and everie landward bairn xxs; and that they shall keep him hous mail frie; and they obless them to pay him yeirlie, in manner above wrytten, xx merks, for reiding in the kirk and raising of the psalmes. And this act to be extendit in ample form.

"Mr William M'Ghie, bailzie.

"Mr George Rutherfurd."

Extracts from the Burgh Records of Kirkcudbright.

17*th October*, 1639.—"The quhilk day the proveist, baillies and councillors of this burgh, being convenit together, all in ane voyce, statutes and ordains that no inhabitant within this burgh, ane or mae, of whatsomever rank, qualitie or degree they be of, presume nor take upon hand, from this tyme furth, under whatsomever culler or pretext, to convocat or assemble themselffes togither at any occasion, except they make dew intimation of the lawfull cause of these meetings to the proveist and baillies of this burgh for the time, and obtein their licence thereto. And that nae tumultis and unlawfull meetings and conventions be maid be them, or any of them, in tyme cuming, quhilk anyways may disturb the quietnes and peace of this burgh; but on the contrair, statutes and ordaines, that the haill inhabitants of this burgh he ready, at all occasiones to concur and assist the magistrates and officers of this burgh for settling of whatsomever tumilts and turbulances that shall happen to fall out, and punishing of the authors and movers thereof. And siclyke statutes and ordaines, that whatsomever person shall hyde themselffes, draw aback, or refuis to concur and assist the said magistrates and officers of the said burgh, readilie, for putting his majesty's laws to execution, within the said burgh, and for settling of whatsomever tumilts shall happen to fall out, shall be repute and halden as vilependers of his majestys authority and laws, and as fosterers and manteiners of the samen tumilts; and shall be punished therefore in their persons and unlawed in their guids, at the arbitriment and discretion of the magistrates and council of this burgh.

Whereupon they ordained ane act, and ordained their clerk and their officer to mak publication heirof at the mercat croce of the said burgh."

1*st May* 1639.—"The quhilk day the proveist, baillies and counsell of this burgh, all with ane consent, statutes and ordaines, that nae inhabitant, burgess or resident within this burgh, presume nor tak upon hand, at any time heirafter to remove out of this toun, upon whatsomever cause or occasion, without liberty of the magistrates first had and obteined thereto; and that they make known to the magistrates the reasson and cause of their going out of the toun, whether it be upon their civil causes or otherways howsoever; but upon the contrair, that they may be ready to attend the magistrates of this burgh, with their armes, whensoever they shall be required, for his majesty's service and for defence of themselffes, their wyffes, childrene and guids, and for saiftie of this burgh frae all invasion whatsomever; and that under the pains of censuring of the disobeyares of this present act in their persones and guids, at the sight and arbitriment of the magistrates and counsell of this burgh, conforme to their qualitie and degrie, and deprivation of their libertie and freedome of this burgh. And ordaines their officers, with touk of drum, to make intimation heirof the next merkat day, that none pretend ignorance heirof."

2*d January* 1641.—"The quhilk day, anent the information sent to us be our commissioner, from ane particular convention of burrowes halden at the burgh of Edinburgh, in November last, and backed with ane missive letter, direct to us be certain noblemen and others of the committee of Estaites of this kingdome, makeing mention, whereas, the Estaites of the kingdome are indebted to the factors in Campheir certain great soumes of money for armes, amunitioun, and other provisions, sent to this kingdome be them for the defence thereof; quhilk soumes were appointed to have been payit to the said factors at this last term. And the Estaites, having been convened and consulted for payment of the said soumes, resolved, that, the haill burrowes of this kingdome should undertake the present advancement of the soums of ane hundred and fiftie thousand guilders, in monie or commodities, to be sent to Holland, for relieff of the said factors. Conforme whereunto, we sent to our said commissioner ane new commission to undertake for our part of the said soume, conforme to the tax roll of the burrowes, upon sufficient securitie for repayment thereof, with interest, sua lang as the samen shall happen to remain unpaid. And we now finding ourselffes unable to advance our part of the foresaid soume, therefore we, all in ane voyce, have determined and concluded, that John Ewart, merchand, ane of the baillies of the said burgh, (wha for the present has the greatest trade in Holland nor any other person in this burgh,) should advance our part of the foirsaid soume to the said factors, either in money or commodities, as he should think best. After the report whereof we declared ourselffes maist willing to give the said John Ewart securitie therefor, as we should be commanded be the committie of Estaites and commissioners of burrowes, upon sufficient securitie to be given to us be the said committie for our relieff of the samen, conforme to ane act maid thereanent betwixt the burrowes and committie of Estaites. Whereupon we ordained ane act to be maid and the samen to be insert and registered in our burrow court books, and has subscryved the samen as follows.—Lykeas, we maid intimation thereof to the said John Ewart, that he should pretend nae ignorance, and ordainit the double heirof to be sent out to the committie of Estaites and commissioners of burrowes, for the exoneration of our diligence in the premises."

6th January, 1641.—"The quhilk day, the judges elect

George Johnston and William Thomson, burgess of this burgh, constables for apprehending of all fugitives, runawayes, maisterless men, and those who travels without ane pass, gif any of these resort to this burgh, and territorie thereof, and for putting of them to the magistrates to be censurit be them, as accords; who have given their aiths de fidele administratione thereuntill."

17th February, 1641.—"The quhilk day, the judges absolves George Callender, present, from the acclaim of William Andersone, whereby he acclaims the said George the soume of twentie punds, quhilk he alledgit he advancet and peyit for Agnes Pauling, and quhilk was imposit upon her for to help the volunteirs of the burgh of Kirkcudbright, and whereof the said George Callender faithfullie promised to sie the said compleiner peyit, in manner alledged in his acclaime; because the said compleiner declaired he went not out as a volunteir; lykeas, the said George Callender promised to make the xxlibs. furthcumane to the weill of the guid cause, and for help of volunteirs of this burgh, when any should happen to go out, and therefore the judges absolves him, as said is."

3d March, 1641.—"Decerns George Meik, present, be the modification of the judges, to pey to Patrick Carsane twelff punds money, in satisfaction to him of the soum of twentie merks money, quhilk the said Parick contributtit and gave to the said George, for helping of his charges when he was ordainit to go out as ane volunteir to the armie, in respect the said George went not to the armie, but onlie went to Edinburgh, and immediatelie thereafter went over to Holland, upon his own affairs. And also, decerns him to delyver to the said Patrick, his saddle that he gave him with his horse to ryde on, as guid as he receavit the samen, with expenses of plea."

2nd June, 1641.—"Ordaines William Fullartoun, proveist, to keip the Committie of War within the burgh, in tyme cuming; and ordaines John Carsane, baillie, to keip them when they hald committies abroad in the countrie."

In October, 1642, a number of those who had joined the covenanters in their expedition into England, as volunteers from the burgh of Kirkcudbright, were admitted free burgesses of the town, as a reward for their services.

7th September, 1643.—"The said day, the bailzies and counsall, anent the chusing of their commanders and officers, within burgh, conforme to the act of the late convocatione, have elected and chosen William Glendonyng, their proveist, Captain; John Carsane, bailzie, Lieutenant; Patrick Carsane, Ensigne; George Callendar, Robert Heuchan, John Clerk and George Meik, Serjands."

4th April, 1644.—"The quhilk day, John Ewart and John Carsane, bailzies, and George Meik, merchand, hes undertane to furnish the toun sufficientlie in arms, to witt—in musket, pyke, sword, match, powder and ball; and shall bring the samen to the said toun, betwixt and the day of May, next to come, and shall pay out the moneys therefor, to witt, the said John Carsane the ane half, and the said John Ewart and George Meik, the other half, equallie between them. For the quhilk, the toun are become actit to give and pay, ilk punds wairing xxvs, for charges and all. And the toune hes condiscendit that the said arms shall be ressavit aff the buyers' handes, within aught dayes after their hame coming, and shall get payment therefor within xx dayes thereafter; and, in caice the toun will not ressave the arms, the merchands shall be liberate of their bargain and permitted to sell the samen to the countrie, or any other who pleis to buy the samen. And the toun hes ordainit their clerk to subscrive thir presents in their behalf."

25th July, 1644.—"The said day John Carsane, bailzie, conforme to the orders given to him anent the buying and bringing home of xxv muskets, had delyverit the samen to the tounship, within their tolbuith. Lykeas, the toun is become actit and oblesit for payment of ilk musket and bandoleer the soum of aught punds for charges and all, and that betwixt and the first day of November next to cume."

29th November, 1644.—"The quhilk day, anent the chusing of the commanders for the toun, as captaines, lieutenants and serjeants, the haill counsall hes condiscendit, that the magistrates shall have the place as Captain; and lykwayes have chosen Patrick Carsane as Lieutenant; William Fullartoun, son to the old proveist, as hand chymer; and John Clark and John Lidderdale as serjants; and ordainit them to cause drill the toun, twice in the weik, upon Monday and Saturday, weeklie. And sicklyke, statutes and ordaines that those absent from the drilling, the first day shall pay half ane dollar, and any other day thereafter six shillings, scots, toties quoties. And sicklyke, the proveist, bailzies and counsall statutes and ordaines, that there be six persones upon the watch everie night, with ane check watch; and ilk persone that shall refrain and be absent from the watch shall pay xxiiijs ilk tyme, toties quoties, and the watch to begin this present day; and the charge of the watch to be upon the lieutenant and other officers above named."

7th December, 1644.—"The quhilk day, it is statuted and ordained, that the inhabitants of this toun keep the watch and ward appointed, and that everie man keep the watch himself, or else to have as ane sufficient man for him, under the paine of xxxs. It is lykewayes, statuted and ordained that the lieutenants and serjeants, in respect that they have the charges of the watch in the setting thereof, that they be exonerated therefrae themselffes."

25th December, 1644.—"The said day, it is statuted and ordainit, that, John Clark, merchand, shall goe to Edinburgh and buy for the toun's use, the amunitione following, viz.—thrie hundred weight of powder, thrie hundred punds of ball, and sex hundred weight of match; and that he goe away upon Monday next."

29th May, 1645.— "The quhilk day, the proveist, baillzies and counsall, taking to their consideratione the great danger may befall the toune, in thir dangerous and troublesome times, throw not keeping of ane strict watch, have for preventing thereof, statuit and ordainit, that there be ten persons upon the watch ilk four and twentie hours, viz.—three at the Meikle Yett, three at the Moat, twa at the Wynd futt, and twa check. The watch to begin this present day, and sua to continue until the samen be dischargit; and that ilk person keep the said watch diligentlie, under the pain of warding of their person for the space of twenty four hours, by and attour the keeping of the watch. The watch to be set at aught hours afternoon, and to continue until that tyme on the morrow. And ordains the check watch to rule the rest of the watch, in setting the watch and bringing them off. As also, that ilk person appear at the ringing the aught hour bell, and in caice they come not betwixt and the ringing out, they are to pay vjs, and in caice they refuis payment, that they be incarcerate us ane delinquent, by and attour the keeping of the watch."

21st July, 1645.—"The quhilk day, in respect that the township hes no money, for the present. to outreik their souldiers, for the expedition in the north, both for arms and viij dayes' provisioun; and withal, considering that William Glendonyng, proveist, hes in his hands, unadvanced, the soum of ix^{xx}lbs, dew be the toun to John Lowrie, merchand, burgess of Edinburgh, conforme to ane band maid be them to the said John, thereupon, for the first month's mantenance. Therefore, because the said John was absent to have resaivit, the said money, at the last tyme the proveist was in Edinburgh, the baillies and counsell hes statuit and ordainit, that the said soum be presentlie employed for outreiking of the said souldiers; and that, with the first convenience, there be ane course taken for payment of the foirsaid soum to the said John Lowrie; that the said William Glendonyng, proveist, be liberate of his band, given be him for the toune thereupon; and for that effect, we presentlie become actit and oblesit therefor, for relief of the said William Glendonyng, sua that he shall incur no danger throw the said band."

"The said day, in respect that Patrick Connell, indwellar within the toune, being nominate ane souldier, for marching to the north, for the toune, and after he gave his promise to John Clerk, ane of the counsall, to march, he nochtheless hes baslie runaway, and contrair all guid example, will not return to his marching. Wherefore, the proveist, baillies and counsall hes statuit and ordainit, that the said Patrick Connell, his wyff, mother and familie be fullie extruded out of the toune, and that they shall have never futting therein, and that his said wyff, mother and familie be extruded the morrow, in the morning. And it is appoynitt, that Alexander Mowat and John Clerk see this act put in execution, and also, that there be intimation maid be the drum, that nae person nor inhabitant within this toun, resett the foirsaid persones, nor any of them, ilk person under the pain of twentie punds."

9th October, 1650.—"The said day, the proveist, baillies and counsall, in reference to the putting of all fencible persons in readiness, upon six hours advertisement; they went about the samen and nominated the hail fencible persons, and took a list of arms; and ordains, that the foirsaid persons be in readiness, upon their advertisement foirsaid; and this was done in respect of the approaching of the Inglishmen to this kingdome."

11*th February*, 1652.—The quhilk day, anent a printed paper sent be the Commissioners of the pretendit Commonwealth of Ingland, from Dalkeith, to the said toun of Kirkcudbright, datit the twentie one day of January last by past; wherein the said commissioners did authorise, as they alledgit, the foirsaid burgh to meit in some convenient place within the said burgh, and then to elect some person of integritie and guid affectione to the weillfare and peace of this kingdome of Scotland, and for that effect to be at Dalkeith upon the sextene day of February instant. The said proveist, bailzies, and counsell, in relatione thereto, having taken the said paper to consideratione, after debaites about the lawfulness or unlawfulness of their sending the said commissioner to the said meiting at Dalkeith, did, upon guid and lawful grounds. being fullie satisfied in their consciences, all with one consent and mynd, resolve, that it was altogether unlawful to give obedience to the said printed paper, and so refuised to send their commissioner to the said meeting at Dalkeith.

1*st January*, 1653.—"The said day, the toune taking to their consideration that there were certain persons burgesses of the said burgh, wha, in 1648, the tyme of the Dukes levie, gave band to the laird of Balmaghie, for payment of the soum of nyne score punds, principal, with certain liequidat expenses and annual rent therein conteinit and restane, of the soum of twentie five hundred punds, for the toun's souldiers, quhilk was payit to Balmaghie; and that the said persons, granters of the band, did give the samen for the weill of the said toune, for the present, and that the said debt aught to be ane common burden upon the said toun. Therefore, the tounship hes condiscendit and are content, that, they shall warrand, relieve, and skaithless keep the foirsaid persons, granters of the said bands, of any payment of the soumes of money therein conteinit, and of all skeath and expenses they may incur therethrow, at all hands, as accords of the law.

Extracts from the Records of the Kirk-Session of Dumfries.

1647.—" By Ordinance of the Synod of Dumfries,— It is to be intimated out of all the pulpits therein, that the persons after-written are excommunicate, and that none reset them, nor resort to them, without licence of Presbyteries or other Kirk Judicatories, upon evidence of necessar and just cause, asked and given, under peril of ecclesiastic censures: they are to say, John Lord Herries; Dame Elizabeth Beaumont, Countess of Nithisdaill; Dame Elizabeth Maxwell, Lady Herries, elder; Elizabeth Maxwell, Ladie Kirkonnell, elder; Helen Maxwell, Lady Mabie, elder; James Maxwell of Kirkonnell, alias, Master of Maxwell; James Lindsay of Auchenskeoch, elder; John Lindsay, his oye; Roger Lindsay of Maynes; Francis Lindsay, his brother; Cuthbert Browne, brother in law to Maynes; Gilbt. Browne of Bakbie; William and Robert Maxwells, brothers to the Laird of Conhaith; ——— Maxwell, sister to Umqle Sir John Maxwell, of Conhaith; Agnes and Janet Maxwells, his daughters; Marion Maxwell, Ladie

Wauchope, elder; Barbara Maxwell, Ladie Auchinfranco; Grizel Gedes, Gudwyfe of Drumcoltran, elder; John Little; John Maxwell, called Captain John; Elspeth Heries, Gudwyfe of Crochmore; Margaret A'Hannay in Kirkgunzeon; Effie Beattie, sometime in Colledge; Wm. Thomson and his Wyfe, in Traqueer; John Maxwell of Mylnestone, alais John of Logane; John Glendynynge, of Parton; Rob. M'Lellan of Nuntone; Elizabeth Young, Relict of Doctor Honyman; Isabel Honyman, daughter to the said Doctor Honyman."

Thursday, 29th *April*.—" The Session give liberty to Mr John Carsan and Mr Cuthbert Cunninghame to speak with the Lord Hereis, notwithstanding he be excommunicate: In respect they both have sundrie business of good—wt his Lordship. And withall they are admonished to refrain their wonted freedom in drinking, with certification if they do in the contrair, they shall make answer.

" Likewise grants the same liberty, to Robt. Newall anent his affairs with Maynes and John Maxwell of Mylnestone."

December 22*nd*., 1648.—" It is ordained that all persons or souldiers, who willingly did embark in the late unlawful engagement, shall forthwith depart this toune and parish, under pain of ecclesiastical censures. It is also ordained that whatsoever delinquent, being convenit as guilty of this offence, and yet obstinately denies the same, shall make public satisfaction, for this denial; bye and attour due repentance for his sin."

Thursday, *June* 21*st*., 1649.—" Anent the humble desire of Mr John Corsane, late provost, to be admitted and received into the covenant: the members of the Session never heretofore being acquainted with the nature of his suit, have found it expedient that he be turned over to the presbytery, as the most fitting and competent judges for clearing his carriage."

"John M'Lean is to be taken into the covenant on Sunday, in relation to his often supplication.

"George Douglas became acted, that if he shall vent any bitter or malignant expressions against the Cause and Covenant of God, or the faithful in the land, he will be willing to embrace the sharpest measure of Kirk censure they can inflict upon him."

Forrester.

Although the Forresters were never, perhaps, proprietors of any very extensive estates in the Stewartry of Kirkcudbright, they nevertheless appear to have been a family of considerable note, and to have ranked and intermarried with many of the most ancient and distinguished families in the district. Several of them, before the Reformation, held offices in the church, and after that period, till about the year 1650, the office of Commissary of the Stewartry was discharged by various members of that family. Indeed the old Register of Deeds, from which so many extracts have been taken in the compilation of the notes of this volume, was written by a Robert Forrester, and in it mention is made of several persons of the name, one of whom was parson of Kirkcormock, about 1580, and another, a Captain Thomas Forrester, who acquired the lands of Kirkland of Kirkcudbright from Maclellan of Bombie, was the commander and principal owner of the first vessel of which we find any notice taken of as belonging to the port of Kirkcudbright. This vessel appears to have been called the Thomas in honor of Sir Thomas Maclellan of Bombie, who at that time, was chief magistrate of the burgh. As it is the first notice we find connected with the shipping of the town, we have subjoined a copy of the deed referring to the vessel, as it is given in the old Register before mentioned.

"At Kirkcudbryt, the xviij day of Merche, the yeir of God Jm Vc lxxxv yeirs, in presence of Robert Forrester, Commissary of Kirkcudbryt.

"The quhilk day comperit in presence of the said commissary the parties underwritten, all personallie, and presentit the discharge underwritten, subscryvit be thame, desyring the same to be insert and registered in the said commissary buiks of the said commissarait, ad perpetuam rei memoriam; quhilk desyre the said commissary thocht ressonable, and therefore hes ordainit and ordains the said discharge to be insert and registered in his said buiks, ad perpetuam rei memoriam; quhairof the tenor follows;—We Mathow Mure, burgess of Glascow, awner of ane quarter of the ship callit the Thomas of Kirkcudbryt, Duncane Sempill, burgess of Dumbartone, awner of half ane quater of the said ship, and Francis Levystoune, burges in Glascow, awner of ane uther half quarter of the said ship; grants us to haif resaivit, be the handes and delyverance of Captain Thomas Forrester, the soume of threttein punds, usual money of Scotland, and that in full satisfactioun, contentment and payment of our said pairts of the said ship, cordage, boit, sails, cabilles, ankeris, and all uthers her ornaments; sauld be us to the said Thomas Forrester, of the quhilk we, and ilk ane of us for our awn pairts, haulds us weill content and thankfullie peyit, as for haill peyment of our said pairts of the said ship, sauld be us to the said Thomas foresaid, and therefore quitclaims, exonerates and discharges the said Thomas of the said soume, in compleit peyment of our pairts of the said ship, his aires, executors and assignayes, for us our aires, executors or assignayes, for now and ever.

And for the mair securitie heirof, we are content and consent, that this our present discharge, be insert and registered in the commissary buiks of Kirkcudbryt, renuncane our awn jurisdiction of Glascow, and all uther priveliges, and subittane us to the jurisdiction of the commissarait of Kirkcudbright in that caice, ad perpetuam rei memoriam. In witness heirof, we have subscryvit thir presents with our hands, as follows;—At Kirkcudbryt, the xviij day of Merche, the yeir of God Jm Vc foure score five yeires, before thir witnesses, Robert Hall, Johne Dicksoun, and Robert Forrester, yungar, burgesses of Kirkcudbryt; Johne A'Sheills, skipper, burgess in Glascow; Johne A'Glukyes, burgess in Glascow.

(sic subscribitur,) Matthow Mure,
Duncane Sempill,
Francis Levystoune,
Johne A'Sheills, as witnes.

Notices regarding the Plague or Seikness.

During the 16th and 17th centuries a dreadful pestilence, which, like the cholera of the present day, had its origin in the east, repeatedly ravaged both the southern and northern parts of Great Britain. During the different visitations made by the plague to this island, it assumed various aspects, and was distinguished by the different appellations of the plague, the pest or pestilence, and the seikness; although it appears that the whole of these were all nearly allied and had their origin from one and the same source—the immense swamps and marshes of India, China and Tartary.

By far the most fatal plagues with which Scotland was visited, were those which occured in the reigns of James VI., Charles I. and Charles II., when such was the terror inspired by the plague, that every bond of society was loosed, and those who had the misfortune to be seized with it were often abandoned by their nearest relations. All intercourse was forbidden to be held between such places as the plague had not yet reached and those in which it was raging,—no large convivial parties were allowed to conveen,—and a fine of ten punds, scots, was imposed on any party who invited more than twelve persons to a marriage, or more than six to the baptism of a child. In many places the mortality was so great, that the formalities of enclosing the body in a coffin and burying in the church-yard were abandoned, and trenches were dug, into which such as died of the plague were thrown. Many superstitious rites were also practised in order to avoid the attacks of this dreaded disease, and in several places the people suspended pieces of raw flesh and bunches of peeled onions on poles, under the impression that whatever infection was in the air would be drawn to and absorbed by them; these after hanging for some time, were taken down, enclosed between two pewder plates and buried with great ceremony. Several of the spots, where the plague was thus interred, and which are generally known by the name of the pest-yards, can still be pointed out in various parts of Scotland, and according to the popular belief, if any of these places were opened, the plague would again break forth, with renewed virulence and desolate the surrounding country.

The following are a few of the notices regarding the plague which occur in the Burgh Records of Kirkcudbright.

12th October, 1586.—" The quhilk day the proveist, bail-zeis and counsell of the said burgh, for certane plessor and guid service done be Jonet Mertene to thame, the tyme of the pest, and for her help, gives and grants to her fyve merks money, being the freidome of Alexander Gordoun, and, for peyment making to her thereof, assigns the said Alexander Gordoun to pey the samen to her; and also, gives and grants to her ane freidome, the next yeir, to sell the samen as she pleiss."

20th April, 1599.—"The quhilk day, the toun understanding the pest being verie ill in Drumfries, and willing nae resorting or traffic be betwixt the said tounes, for feir thairof, statuets and ordains, that the haill inhabitants within this burgh, that raises reik therein, shall keip watche in their awn persoune, or else, ane as sufficient in their steid, at the sicht of the bailzeis, under the paine of xls. ilk fault, and tinsall of their freidome. And that nane haunt or gang to Drumfries, or benethe the Watter of Urr, or reset any benethe said watter, under the said paine, without consent of the bailzeis."

9th December, 1644.—"The quhilk day, the proveist, bailzies and counsall, taking to their consideration, the immenent danger that the toune is in, throw in-coming of beggars and others, at sundrie parts in this toune, wherethrow the toune, in respect that the plague is spreiding in this kingdome, and there is no certaintie from what place the samen may come, for preventing whereof, it is statuit and ordainit, there be two ports biggit up sufficientlie with timber and staines, ane at the meikle yett, and ane other at the wynd futt, with all possible diligence, and that workmen be presentlie set on futt thereto. And also, it is statuit and ordainit, that all heritors, frae the yaird futt perteiening to umqle Robert Glendonyng, notar, to the moat well, big up their yaird futtes, sufficientlie, with staine and other materials necessar, and that they enter thereto presentlie, and appoints William Halyday, treasurer, to oversee this work above-noted sufficientlie done."

23rd November, 1648.—"The quhilk day the proveist, bailzies and counsell, being certainlie informit of the spreiding of the dangerous infection of the pestilence, in the parochines of Tungland and Keltoune, have for the toun's safetie, statuit and ordainit, that nane, inhabitants within the said parochines, shall resort to this burgh, in tyme cuming, untill it pleise God, in his infinite mercie, to remove the said visitation from them; except there be ane pass under the subscriptione of the laird of Bargattone and the guidman of Barcaple, for the said parochen of Tungland, and ane under the subscription of Mr James Fergusson, Minister, at Keltoune, for the said parochen of Keltoune. And that nane of the inhabitants of this burgh have any correspondence with any of the said parocheners, in time cuming, as they will be answerable upon their perill."

GLOSSARY.

Adebted, *indebted.*
Aither, *either.*
Anent, *concerning.*
Attour, *over and above.*
Awn, *own.*
Awing, *owing.*

Battered, *pasted.*
Beirane, *bearing.*
Belliblinde, *blindman's buff.*
Bringane, *bringing.*

Committane, *committing.*
Croce, *cross.*
Culyeon, *a poltroon.*

Delaitor, *informer.*
Distrenzie, *distrain.*

Factorie, *agency.*
Failzie, *fail.*
Flighted, *dismantled or rendered untenable.*
Forneishing, *furnishing.*
Fourfaulding, *forfeited.*
Freithe, *liberate or free.*

Gorgets, *a sort of pillory, the criminal being fastened to a wall by an iron collar which was put round the neck.*
Grasounes, *grass lands.*

Hoggers, *footless stockings.*
Hounsell, *a small meal taken in the morning.*

Inglis, *English.*

Knock, *clock.*
Kyndlie, *natural.*
Kythe, *manifest.*

Mae, *more.*
Mail-frie, rent-free.
Meane, to, *to make mention of, to show or make known distinctly.*
Mercat, *market.*
Messor, *measure.*

Nawayes, *nowise.*

Obleis, *oblige.*
Outreiked, *rigged out, furnished.*
Owand, *owing.*
Owile, *oval.*
Oy, *grandson.*

Parochen, *parish.*
Piecess, *communion cups.*
Plessor, *pleasure.*
Prescryvit, *prescribed.*

Quha, *who.*
Quhat, *what.*
Quhatsumever, *whatever.*
Quhilk, *which.*
Quhom, *whom.*
Quitclaims, *frees from any claim.*

Redrest, *remaininy portion.*
Refuisset, *refused.*
Rewle, *rule.*
Rexane, *ruling, or exercising authority.*
Runawayes, *deserters.*

Samen, *same.*
Scarcietie, *scarcity.*
Schawing, *showing.*
Schuil, *school.*
Soccarer, *one who holds his lands by soccage—a tenant subject to restrictions and bound to perform certain services.*
Submittane, *submitting.*
Subscryvit, *subscribed.*
Suithe, *truth.*

Table chakers, *draught board.*
Thir, *these.*
Tocher, *dowery.*

Yetts, *gates.*

Zeard, *yard, garden.*

Kirkcudbright:—Printed by J. Nicholson.

Lightning Source UK Ltd.
Milton Keynes UK
09 March 2011

168989UK00001B/57/A

9 780548 291092